Global Change and Local Places

Global Change and Local Places explores the ways people and biota contribute to climate change in four localities of the United States. The volume summarizes the findings of the Global Change and Local Places (GCLP) project initiated by the Association of American Geographers (AAG) to develop methods to determine: how localities contribute to global change; how those contributions change over time; what forces drive such changes; what controls local residents exercise over such driving forces; and how locally initiated efforts at mitigation and adaptation can be promoted. The sources and driving forces for greenhouse gas emissions vary widely among the four research sites, as do the possibilities and propensities to mitigate emissions and adapt to the local changes global warming could bring. Policy makers and legislators will be unable to address human-induced climate change effectively without the insights that are revealed by examining and understanding the daily routines that are simultaneously the sources of climate change and the keys to reducing its severity and coping with its effects.

Global Change and Local Places

Estimating, Understanding, and Reducing Greenhouse Gases

BY

Association of American Geographers Global Change
and Local Places Research Team

CAMBRIDGE
UNIVERSITY PRESS

PUBLISHED BY THE PRESS SYNDICATE OF THE UNIVERSITY OF CAMBRIDGE
The Pitt Building, Trumpington Street, Cambridge, United Kingdom

CAMBRIDGE UNIVERSITY PRESS
The Edinburgh Building, Cambridge CB2 2RU, UK
40 West 20th Street, New York, NY 10011-4211, USA
477 Williamstown Road, Port Melbourne, VIC 3207, Australia
Ruiz de Alarcón 13, 28014 Madrid, Spain
Dock House, The Waterfront, Cape Town 8001, South Africa

http://www.cambridge.org

pⱽ

First published 2003

Printed in the United Kingdom at the University Press, Cambridge

Typeface Times 10/13 pt and Helvetica   *System* LATEX 2$_\varepsilon$ [TB]

A catalog record for this book is available from the British Library

Library of Congress Cataloging in Publication data

Abler, Ronald.
Global change and local places : estimating, understanding, and reducing greenhouse gases / Ronald F. Abler;
edited by Association of American Geographers Global Change and Local Places Research Group.
p. cm.
Includes bibliographical references and index.
ISBN 0 521 80950 9
1. Greenhouse gases–Environmental aspects.   2. Air quality management.   3. Climatic changes.
I. Association of American Geographers. Global Change and Local Places Research Group.   II. Title.
TD885.5.G73 A25   2003
363.738′74–dc21   2002035014

ISBN 0 521 80950 9 hardback

Contents

Contributors

Ronald F. Abler served as Executive Director of the Association of American Geographers (AAG) from 1989 through 2002 and is currently Secretary General and Treasurer of the International Geographical Union. Prior to joining the AAG, he taught in the Department of Geography at the Pennsylvania State University. From 1984 to 1988, Abler was Director of the Geography and Regional Science Program at the National Science Foundation in Washington, D.C. His research has been devoted to the geography of intercommunications technologies. He has examined the ways different societies have used such media at different times and places.

David P. Angel is Associate Professor of Geography and Dean of Graduate Studies and Research at Clark University. An economic geographer by training, Dr Angel's current research focuses on issues of industrial change and the environment. His research includes work conducted for the United States Department of Commerce, the United States Agency for International Development, and the MacArthur Foundation. Dr Angel holds the Laskoff Professorship in Economics, Technology and the Environment at Clark University.

Samuel A. Aryeetey-Attoh is Professor of Geography and Chair of the Department of Geography and Planning at the University of Toledo. Dr Attoh specializes in urban geography and urban and regional planning. The United States Department of Housing and Urban Development, the Kellogg Foundation, and the Ohio Board of Regents Urban University Program have sponsored his current research on urban growth management and community sustainability.

Robert Corell is Senior Research Fellow at Harvard University's John F. Kennedy School of Government and a Senior Fellow of the American Meteorological Society.

Susan L. Cutter is Carolina Distinguished Professor at the University of South Carolina where she also serves as the Director of the Hazards Research Laboratory. She was President of the Association of American Geographers in 2000–2001. Her research examines human responses to environmental risks and hazards. Cutter's work has advanced the understanding of decision making and public policy changes in the hazards and global change arenas. She was elected a Fellow of the American Association for the Advancement of Science in 1999 in recognition of her contributions in the field of environmental hazards and assessment and for imparting understanding of environmental risks to communities and individuals.

Jennifer DeHart is a doctoral candidate in the Department of Geography at the University of North Carolina at Chapel Hill and Assistant Professor in the Environmental Science Department at Allegheny College. Her research has focused on the inventorying and reduction of greenhouse gas emissions at various spatial scales, and on watershed regulation. She served as research assistant for the Global Change and Local Places project, investigating the driving forces of greenhouse gas emissions and local capacities to respond to global warming.

Andrea S. Denny received her MS degree in Geography from the Pennsylvania State University in 1999. Her graduate research focused on greenhouse gas emissions and mitigation in Central Pennsylvania. She is currently employed as an Environmental Protection Specialist in the State and Local Climate Change Program at the United States Environmental Protection Agency.

William E. Easterling is Professor of Geography and Earth System Science at The Pennsylvania State University. He directs the newly formed Center for Advanced Carbon Cycle Research and Education. Previous posts include Director of the Great Plains Regional Center for Global Environmental Change at the University of Nebraska-Lincoln and Fellow in the Climate Resources Program at Resources for the Future. His research has been on the role of scale in modeling agricultural response to anthropogenic climate change. He has served on numerous scientific advisory committees and recently chaired a National Research Council panel that published *Making Climate Forecasts Matter*. He was the convening lead author for the agriculture section in the Third Assessment Report of the Intergovernmental Panel on Climate Change.

Douglas G. Goodin is Associate Professor of Geography at Kansas State University. A member of the Kansas State University faculty since 1993, his research interests are in climatology, landscape ecology, and remote sensing. Currently, his work focuses on the interaction of climate, physiography, and disturbance on the evolution of spatial complexity in tallgrass prairie vegetation.

John Harrington Jr. is Professor and Head of the Department of Geography at Kansas State University. His research focuses on climatology, remote sensing, and natural resource applications of geographic information systems. Major long-term research projects have involved remote sensing of suspended sediments in south-central United States lakes and reservoirs and the establishment of remote sensing and GIS analytical capabilities for the government of Niger. Prior to 1994, Harrington held academic positions at the University of Oklahoma, the University of Nebraska, New Mexico State University, and Indiana State University.

Lisa M. B. Harrington is Associate Professor of Geography at Kansas State University, where she has been since 1994. Her research specialties are environmental use and management, including rural land use, human–environment relations, and biotic resources. Much of her work has been based in the Pacific Northwest, along with research related to the Great Plains. She previously held faculty positions at Western Washington, Central Michigan, and Eastern Illinois universities.

Arleen A. Hill is a Ph.D. candidate in the Department of Geography and research assistant in the Hazards Research Laboratory at the University of South Carolina. She received her B.S. from Mary Washington College and her M.S. from the University of South Carolina. Her research focuses on natural hazards, specifically the role scientific uncertainty plays in individual responses to environmental threats.

David G. Howard is Adjunct Professor in the Department of Geography and Planning at The University of Toledo. His research interests are in geographic education and Lake Erie environmental issues.

Sylvia-Linda Kaktins is a doctoral candidate at Kansas State University focusing on sustainability and evaluating quality of life in rural communities. As a rural geographer, her primary interests have been in environmental and resources issues. She served as a research assistant for the Global Change and Local Places project investigating the driving forces behind greenhouse gas emissions, local perceptions of global warming, and the use of analogs to determine how communities will respond to global warming.

Robert W. Kates is University Professor Emeritus at Brown University. A Past President of the Association of American Geographers, he is Executive Editor of *Environment* magazine, co-chair of Overcoming Hunger in the 1990s, Distinguished Scientist at the George Perkins Marsh Institute at Clark University, and Faculty Associate of the College of the Atlantic. From 1986 to 1992, Kates directed the Alan Shawn Feinstein World Hunger Program at Brown University. His current research and professional interests include: the prevalence and persistence of hunger; long-term population dynamics; sustainability of the biosphere; climate impact assessment; and theory of the human environment. Kates is a recipient of the 1991 National Medal of Science awarded by the President of the United States, the MacArthur Prize Fellowship (1981–85), and Honors from the Association of American Geographers. He is a member of the National Academy of Sciences, the American Academy of Arts and Sciences, and a fellow of the American Association for the Advancement of Science.

C. Gregory Knight is Professor of Geography at the Pennsylvania State University. He served as Head of the Department of Geography from 1982 to 1989 and Vice Provost of the University from 1989 to 1993. From 1996 through 2002, he was founding Director of the Center for Integrated Regional Assessment, a National Science Foundation sponsored project at Penn State. His research in the field of global change concerns climate change impacts at local and regional levels, with focus on water resources and southeastern Europe.

David E. Kromm is Professor Emeritus of Geography at Kansas State University, where he taught since 1957. His research has included the role of forestry in economic development in Michigan, public response to air pollution in Slovenia, regional water management in Britain, and irrigation efficiency in Alberta. Since 1979 he has been a principal investigator in several sponsored studies examining various aspects of groundwater depletion on the American High Plains.

Steven Lachman is a doctoral candidate in Geography at Penn State University (Ph.D. expected Spring 2003). His dissertation compares the environmental sustainability of two communities, State College, PA, and Burlington, VT, and how their sustainability is influenced by the globalist capitalist economy. Lachman also has a Master's degree in Geography from Penn State and a Law degree from Vermont Law School. From 1988 to 1997, he worked as an attorney for the Pennsylvania Department of Environmental Protection. He is also a trained mediator.

Peter S. Lindquist is Associate Professor in the Department of Geography and Planning at The University of Toledo. Prior to this appointment he served on the faculty in the Department of Geography–Geology at Illinois State University from 1990 to 1996. His primary research interests are in geographic information science, location theory, and urban transportation planning.

Neal G. Lineback is Professor of Geography and former chair of the Department of Geography and Planning at Appalachian State University. From 1969 until 1987, he taught and served as chair of the Department of Geography at the University of Alabama. He is originator and author of *Geography in the News*, a weekly newspaper column, syndicated for the past fourteen years. He is active in research on greenhouse gas emissions and strategies for their mitigation.

Michael W. Mayfield is Professor and a former chair of the Department of Geography and Planning at Appalachian State University. He has previously taught at the University of Idaho and the University of North Carolina at Greensboro. His research interests are in global change and surface hydrology. His global change research has focused on developing methods for estimating greenhouse gas emissions and carbon sequestration in forests at the local level.

Jerry T. Mitchell is an Assistant Professor of Geography at Bloomsburg University of Pennsylvania. His research has focused primarily on the social and cultural dimensions of hazards, primarily with respect to environmental justice, hazard vulnerability, and hazard perception. He received his Ph.D. from the University of South Carolina in 1998.

William A. Muraco is Research Professor & Professor Emeritus at the University of Toledo. His research interests are Economic Geography and Health Planning.

Colin Polsky is a NOAA/UCAR Climate and Global Change Postdoctoral Fellow with the Research and Assessment Systems for Sustainability program at the Belfer Center for Science and International Affairs, Harvard University. He is trained as a geographer, with undergraduate majors in mathematics, humanities and French. Colin's principal research area is the human dimensions of global environmental change, emphasizing the statistical analysis of vulnerability to climate change. His Ph.D. research focused on regional assessments of climate change impacts in agriculture. At Harvard Colin is promoting statistical techniques for assessing socio-ecological vulnerability to environmental changes in two regions, the Arctic and the United States Great Plains. Dr Polsky joined the faculty of the Geography Department at Clark University in 2003.

Neil Reid is Associate Professor in the Department of Geography and Planning at the University of Toledo. His primary research interests are in the spatial dynamics of economic restructuring in the United States. Much of his research has focused upon the role of Japanese direct investment in this economic restructuring process.

Audrey Reynolds received her Bachelor's degree at The Johns Hopkins University in 1992 and earned a Master's degree in Geography at Appalachian State University in 1998. She was deeply involved in managing North Carolina's greenhouse gas emissions data base before moving to the University of Texas in Austin in 1999. Her research interests include both the physical and human patterns involved in environmental change.

Robin Shudak works in the Environmental Protection Agency's Energy Star Program, where she is Special Assistant to the Program Director. She received her Bachelor's degree in Geography from Texas A&M University and her Master's degree in Geography from Penn State. Her research interests include both the physical and human dimensions of environmental variation and change.

Stephen E. White is Interim Dean of the College of Arts and Sciences at Kansas State University. He was Head of the Department of Geography from 1979 to 1987 and again from 1994 to 1997, and has served as Associate Dean of the College of Arts and Sciences since 1997. His research includes internal migration and population change in central Appalachia and the Great Plains. He has conducted several long-term, sponsored research projects on the High Plains – Ogallala Aquifer region, focusing on public perceptions of groundwater conservation, adoption of water-saving technologies, and institutional responses to groundwater depletion.

Thomas J. Wilbanks is Corporate Research Fellow and Leader of Global Change and Developing Country Programs at Oak Ridge National Laboratory (ORNL). A Past President of the Association of American Geographers, he has chaired the United States National

Committee for the International Geographical Union and the 1995–97 National Academy of Science – National Research Council Committee on Rediscovering Geography: New Directions for a New Century. He received AAG Honors in 1986, the Distinguished Geography Educator Award of the National Geographic Society in 1993, and the Anderson Medal of Honor in Applied Geography in 1995. The programs Wilbanks oversees at ORNL have been involved in energy and environmental projects in more than 40 developing countries in the past decade. In recent years he has been involved in the United States National Assessment of Possible Consequences of Climate Variability and Change, Intergovernmental Panel on Climate Change Working Group II (Impacts, Adaptation, and Vulnerability) and the Millennium Ecosystem Assessment.

Brent Yarnal is Professor of Geography and Director of the Center for Integrated Assessment at the Pennsylvania State University. He is co-founder and Past Chair of the Association of American Geographer's Human Dimensions of Global Change Specialty Group and past chair of the Association's Climate Specialty Group. He co-edited the journal *Climate Research* from 1996 through 2001. His research focuses on the local and regional dimensions of global environmental change.

Foreword

Robert W. Corell

Think globally and act locally has been a slogan of the environmental movement for many years. The idea spawned an awareness of changes in our environment that have led individuals and communities across the nation and around the world to undertake recycling, energy conservation measures, and other actions that have helped increase our stewardship. While many of the observed changes could be seen in our immediate environment and surrounding our communities, there was a growing recognition that there were other changes on a grander scale: rapid rates of change in global average atmospheric temperatures and the extremes in weather, increased precipitation in some regions while major droughts occurred in others, ozone depletion and greenhouse gas increases, sea-level rise, and other measures of the state of the planet's life support systems. Although the life-enabling envelope that surrounds the planet has been changing forever on many time scales, the acceleration in rates of change during the past century or two has not been seen for 10,000 years or more in some cases, and others not for millions of years. While the environmental movement was a product of the 1960s that led to recycling and other measures of environmental stewardship, the early 1980s brought a stark recognition that there were other changes which are global in reach, that could have profound long-term implications, and which could be anthropogenic in origin. It was this new 'global change' perspective that led to a remarkable two-decade period of primarily global-scale research in which tens of thousands of scientists were engaged and in which billions of dollars were invested to understand more fully the processes that control these global-scale changes as well as their potential for the future. By the late 1980s, governments, other organizations, the private sector, and the public sought to develop complex and often politically charged agreements that would establish policies, implement practices, and develop technologies to limit the long-term consequences of these changes.

By the late 1990s, the sciences of global change had developed a first-order and, in some cases, fairly complete understanding of the global trends, averages or means, and patterns of change. More importantly, it became disturbingly clear that virtually every parameter observed or modeled during the global-scale research programs varied widely on both temporal and geographical scales. In short, the Earth system's behavior is lumpy and often sharply discontinuous. For example, the ozone depletions in the Antarctic occur annually and result in a massive but relatively simple hole in the ozone layer over the Antarctic continent; in the Arctic, depletions are less predictable in time and tend to be spotty. Another example is temperature variations across the Arctic over the past several decades, which show Greenland cooling consistently over most of its extent while Alaska has been warming dramatically. These differences are evident in both the natural systems of the planet and in the ways humans behave and inhabit the Earth. The consequences of these changes, whether from increases in greenhouse gases or from the urbanization and changing demographics of our societies, are profoundly important. What is now missing is the scientific capacity to move beyond the knowledge and understandings obtained at global scales and extend those insights to particular places. This reorientation toward local places (or what is now often referred to as *place-based* science) is simple in construct but profoundly challenging to implement. At global scales, such scientific questions as the following are being asked:[1] (1) What are the critical thresholds, bottlenecks and switches in the Earth system? (2) What are the Earth's major dynamical patterns, teleconnections, and feedback loops? (3) What are the characteristic regimes and time scales of Earth's natural variability? (4) What are the anthropogenic disturbance regimes that matter at Earth scale, and how do they interact with abrupt and extreme events? (5) Which are the most vulnerable regions undergoing global change? (6) What are the human carrying capacities of the Earth under various assumptions and values? and (7) many others of similar structure and substance. While these questions are profoundly important, taken alone, they will not be able to address the issues of changes that occur at local levels, the very levels at which global changes are experienced by communities and individuals and where adaptations are most likely to take place.

   Place-based research is clearly a local perspective but it must be nested in the global system, where scale matters, and where scale interactions govern both upward to the more regional and global levels, and downward to the most personal levels. Global change expressed at local places, as pioneered by this remarkable book, is premised on a simple observation that the stresses arising from 'severe environmental degradation can be difficult to untangle from one another; they are shaped by the physical, ecological, and social interactions at particular places, that is locales or regions. Developing an integrated and place-based understanding of such threats and the options for dealing with them is a central challenge for promoting a transition toward sustainability.'[2] The questions raised at the local sites, as studied throughout the chapters in this analysis of global change and local places, are markedly different from those raised at the global Earth scale. The questions that need to be addressed for local scale insights are very well represented by the three fundamental questions addressed in all four of the study sites examined in this book. They include:

• What have been the magnitudes and trajectories of local changes in greenhouse gas emissions and their sources and sinks since 1970, and what are they likely to be through 2020?
• What forces have driven these emission trajectories in the past, and what forces are likely to drive them in the future? and

---

[1] These questions were taken from *Global Change and the Earth System: A Planet under Pressure*, IGBP Science Series 4, ISSN 1650-7770(2001).
[2] Taken from a report of the National Academy of Sciences and its National Research Council, *Our Common Journey, A Transition Toward Sustainability* (1999).

- What local capabilities exist to moderate changes in driving forces, are there local propensities to undertake such adaptive behavior, and how might possible policy mandates affect adaptation?

These questions have guided the analyses in this book and are likely to guide others that follow. In short, the lessons learned from this study are summarized well and succinctly:

Simply stated, the Global Change and Local Places project shows that local knowledge is important, although not for everything. The beguiling slogan *think globally and act locally* does not suffice for climate change, its causes, and its consequences because global or even national knowledge averages too many distinctive local trajectories of greenhouse gas emissions and their driving forces, overlooking positive mitigation opportunities and making local actions more difficult. Local knowledge, however, is also an incomplete basis for action, since most decisions that could result in major local emission reductions will not be made locally. Revising the familiar slogan to recognize how the local relates to global calls for three insights: *make the global local; look beyond the local; and act globally in order to act locally.*

One might also speculate that if scientific research on major global change and climate questions becomes increasingly place-based and set in a global framework, then policy-making and action may increasingly be more local and regional, but similarly nested in global protocols and conventions. For example, while the United States has had major difficulties in accepting the targets and timetables of the Kyoto Protocol for Climate Change, the Conference of New England Governors and eastern Canada's provincial premiers have pledged to reduce greenhouse gases 10% below 1990 levels on the time lines set by the Kyoto protocol. These reductions will be followed by other local policy actions and stewardship practices by both local governments and the private sector. This book provides new insights and guidelines for *thinking globally and acting locally*. Thinking globally now evolves from a profoundly richer scientific understanding than was possible a generation ago, and acting locally over the decades ahead will increasingly contribute toward achieving a more sustainable future for the planet and its peoples.

This is a remarkable and important book, the first of its kind to offer such depth of analysis and perspective at local levels. It marks a pathway for future studies by others and provides a model methodology for analyses. It is a privilege to introduce to you this book by a gifted set of scientists and authors. Congratulations too, to the Association of American Geographers for enabling the analysis that has lead to this important new treatise on *Global Change and Local Places*.

Preface

Ronald F. Abler

Global Change and Local Places explores the ways people and biota contribute to climate change in localities of the United States. The volume summarizes the findings of the Global Change and Local Places project initiated by the Association of American Geographers (AAG). The Global Change and Local Places project was underwritten by the National Aeronautics and Space Administration's Office of Earth Science to develop methods to answer such questions as: how do local places contribute to global change; how do those contributions change over time; what forces drive such changes; what controls do local residents exercise over such driving forces; and how can locally initiated efforts at mitigation and adaptation be promoted?

Four study areas were selected for environmental and economic diversity and for the local knowledge, scientific competence, and commitment to long-term study on the part of the four academic institutions that participated in the project: Kansas State University (Manhattan, KS), Appalachian State University (Boone, NC), the University of Toledo (Toledo, OH), and the Pennsylvania State University (State College, PA). The four sites offer, respectively, intensive irrigated crop and beef production in Kansas, a mix of forestry and manufacturing in a rapidly growing area of North Carolina, a restructured belt of heavy industry in Ohio, and extensive open-pit coal mining and lime production in Pennsylvania. The sites are local, but not small. They are the equivalent of 1° of equatorial latitude and longitude, or about 12,300 square kilometers (4750 square miles), or about the size of the state of Connecticut. Study area boundaries were fixed along county lines. The one degree size was chosen so the study areas could in time be linked to the standard grids of climate change models, which then used five to ten degree grids.

Each study area research team has assessed the changes from 1970 to 2020 in human activities that alter greenhouse gas emissions and uptakes, greenhouse aerosols, and the forces driving changes in those human activities. With these emissions inventories and driving force assessments in hand, the research teams surveyed the major emitters and a sample of households in each study area to evaluate their knowledge, capacity, and willingness to mitigate emissions and to adapt to climate change.

The findings from the four study areas provide the foundation for an examination of the ways emissions and their driving forces vary among study areas, through time in the same study areas, what local residents know about driving forces, how they have dealt with similar changes and forces in the past, and how much they worry about climate change and reducing their greenhouse gas emissions. Those findings in turn provide a basis for assessing the opportunities and challenges local residents will encounter in meeting international commitments to reduce greenhouse gases. *Global Change and Local Places* concludes with the Global Change and Local Places research team's considered observations on improving the capacity to study global change in localities worldwide, and with a summary of the Global Change and Local Places project's major findings and their implications.

Acknowledgments

The Association of American Geographers Global Change and Local Places Research Team gratefully acknowledges the support of the United States National Aeronautics and Space Administration (NASA) for the research reported in this volume. Funding was provided by the NASA Mission to Planet Earth Program (subsequently designated Destination Earth) through a grant to the Association of American Geographers. The United States National Science Foundation provided funding (Award NSF 9310459) for a Workshop on Geographical Contributions to Global Change Research in Atlanta, GA, in April 1993, where the early plans for the Global Change and Local Places project were formulated.

The Publisher has used its best endeavors to ensure that the URLs for external websites referred to in this book are correct and active at the time of going to press. However, the publisher has no responsibility for the websites and can make no guarantee that a site will remain live or that the content is or will remain appropriate.

PART ONE

Global change in local places

1

A grand query: how scale matters in global change research

Robert W. Kates and Thomas J. Wilbanks

Grand queries are fundamental questions that transcend the form and substance of individual sciences; they often appear simultaneously in many disciplines. A recurring grand query focuses on scale: how to relate universals to particulars, wholes to parts, macro-processes to micro-behavior, and global to local. Biologists ponder the linkages among molecules, cells, and organisms; ecologists among patches, ecosystems, and biomes; economists among firms, industries, and economies; and geographers among places, regions, and Earth (Rediscovering Geography Committee 1997: 95–102; Alexander *et al.* 1987; Holling 1992; Levin 1992; Meyer and Turner 1998; Meyer *et al.* 1992; Turner, M. G. *et al.* 1993). Scientists in many disciplines worry about non-linear processes and complexity: whether understanding its components can explain the properties of a large system (Gallagher and Appenzeller 1999). Or the reverse, as in the case of global climate change: can the rapidly accruing understanding of the large Earth system inform the ways people and biota in particular places alter climate and in turn are affected by climate change?

This chapter places the Global Change and Local Places project in the context of a grand query: how *scale* matters in global climate change. It examines the scale at which global change and responses to it take place, and how well the current scales of science and policy match the current scales at which changes are engendered. This analysis is rooted in a simple causal chain of human-induced climate change and a brief summary of the current state of scientific knowledge for each link in that chain. The analysis draws heavily upon the third assessment report of the Intergovernmental Panel on Climate Change (Houghton *et al.* 2001; McCarthy *et al.* 2001; Davidson *et al.* 2001).

The causal chain consists of six links: (1) driving forces, (2) human activities, (3) radiative forcing, (4) climate change, (5) impacts, and (6) responses. The Global Change and Local Places team examined the geographic scale of each link and asked how well current scales of assessment – observation, research, and policy – match the scales of each of the six processes. For some links, serious scale mismatches exist between processes and assessments, mismatches that the global change research community increasingly recognizes. The problems such scale disjunctures cause can be addressed both by moving current approaches downscale and by employing the bottom-up approach taken in the Global Change and Local Places project and in similar studies of global change.

How scale matters

Wilbanks and Kates (1999) have suggested two sets of arguments as to how scale affects global climate change. The first set concerns the way the world works. Human-induced climate change arises from interactions between the different domains of nature and society, each composed of many systems operating at different scales in space and through time, resulting in mismatches in scale between causes and consequences. For example, many social scientists seek to understand relationships between *agency* (intentional human action) and *structure* (institutions and other regularized relationships within which human action takes place). The scale of agency is almost always more localized than that of structure.[1] When global structure and local agency interact across different domains, on different time scales and over different areas, the resulting relationships are neither easily understood nor readily predictable.[2]

The second set of arguments is rooted in the practice of science. Current ways of relating global climate change to localities are top-down: from the global toward the local. They begin with climate change scenarios derived from global models, even though those models have little regional or local specificity. But at global scales, understanding the complex interactions among the environmental, economic, and social processes that drive change often seems intractable (Cox 1997).[3] Place-based research, well-grounded in local experience, offers a more tractable alternative for tracing these complex relationships. Though locality studies may be more tractable, however, they are also less generalizable. Where possible therefore, case studies should constitute natural experiments carefully chosen for comparability and investigated by using a common study protocol.[4]

Small study areas offer variety as well as tractability. The variance from a sample of small geographic areas will likely be greater than the variance from a sample of large areas (Figure 1.1). The greater variety in processes and relationships at local scale represents an opportunity for learning about causes and consequences of global climate change that

---

[1] For example, consider the range of human responses to natural and technological hazards. Most major decisions are made locally (Cutter 1993), but within larger structures that *mandate* some actions by law, regulation or court order, *encourage* some actions through persuasion or incentives, and *inform* those creating risk (who may voluntarily reduce it) and those suffering risk, who may learn to tolerate the hazard (Kasperson *et al.* 1985; Palm 1990; Cutter 1993; Hewitt 1997). Other literature reinforces this impression, considering evidence about the scale of human-ecological self-determination (Wilbanks 1994) and scale and consensual decision-making regarding the use of technology (Wilbanks 1984; Aronson and Stern 1984; Chapter 7).

[2] Modelers such as Holling (1995: 27) identified a few cases of managed ecosystems (boreal forests, boreal prairies, and pelagic systems) where the relevant scales of sizes and speeds and their interactions are well understood. But in regions where human activities predominate, interactions are more complex; a study of nine 'regions at risk' where large subnational zones are undergoing great environmental stress found interactions to be highly diverse and complex (Kasperson *et al.* 1995).

[3] Root and Schneider (1995) suggest *strategic cyclical scaling* for analyzing interactions among processes operating at different climatic and ecological scales. Strategic cyclical scaling would involve continuing cycling of studies between large-scale associations that suggest small-scale investigations in order to test the causes and driving forces of the large-scale patterns.

[4] For examples, see the case studies of natural hazards by White (1974), population density and agriculture in Africa by Turner, B. L. *et al.* (1993), and poverty and environment by Kates and Haarmann (1992).

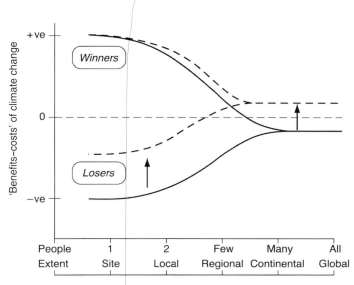

Figure 1.1 Scale-dependent distribution of impacts of climate change (adapted from Environment Canada).

often are obscured when behavior is averaged over larger areas.[5] Finally, in many situations researchers looking at an issue from a global perspective come to conclusions different from those reached by investigators looking at that same issue from a local perspective.[6] Focusing exclusively at a specific scale can lead to conclusions that are highly dependent on the information collected, the parties seen as influential, and the processes that operate at that scale; information, actors, and processes that operate at other scales may be overlooked.[7]

Climate change: causes and consequences

Central to scale considerations in climate change is the global greenhouse effect whereby natural and human-induced greenhouse gas emissions affect radiative forcing of the climate. Solar radiation is the essential source of life on Earth, providing heat and light and powering the movements of air and water that humans experience as climate. Life on Earth has

[5] For example, persistent decadal fluctuations of greater than average temperature (up to 1 °C annually or 2 °C seasonally) and precipitation (approximately 10% annually) have occurred in most areas of the United States during the period of modern records. These natural variations that mimic what might occur as a result of global warming have been identified for all climate divisions of the United States (Karl and Riebsame 1984). In one study, such fluctuations were used to study the impacts of possible global warming on the runoff portion of the hydrologic cycle (Karl and Riebsame 1989).

[6] For example, macro-scale analysis of climate change impacts on agriculture finds little net loss in productivity; one region's gains offset another region's losses, especially with carbon dioxide fertilization and modest levels of adaptation (Fischer et al. 1994). Micro-level studies identify developing country smallholder agriculturists, pastoralists, wage laborers, the urban poor, refugees, and other destitute groups as especially vulnerable (Bohle et al. 1994).

[7] For instance, Openshaw and Taylor (1979) have demonstrated that simply changing the scale at which data are gathered can change the correlation between variables virtually from +1 to −1.

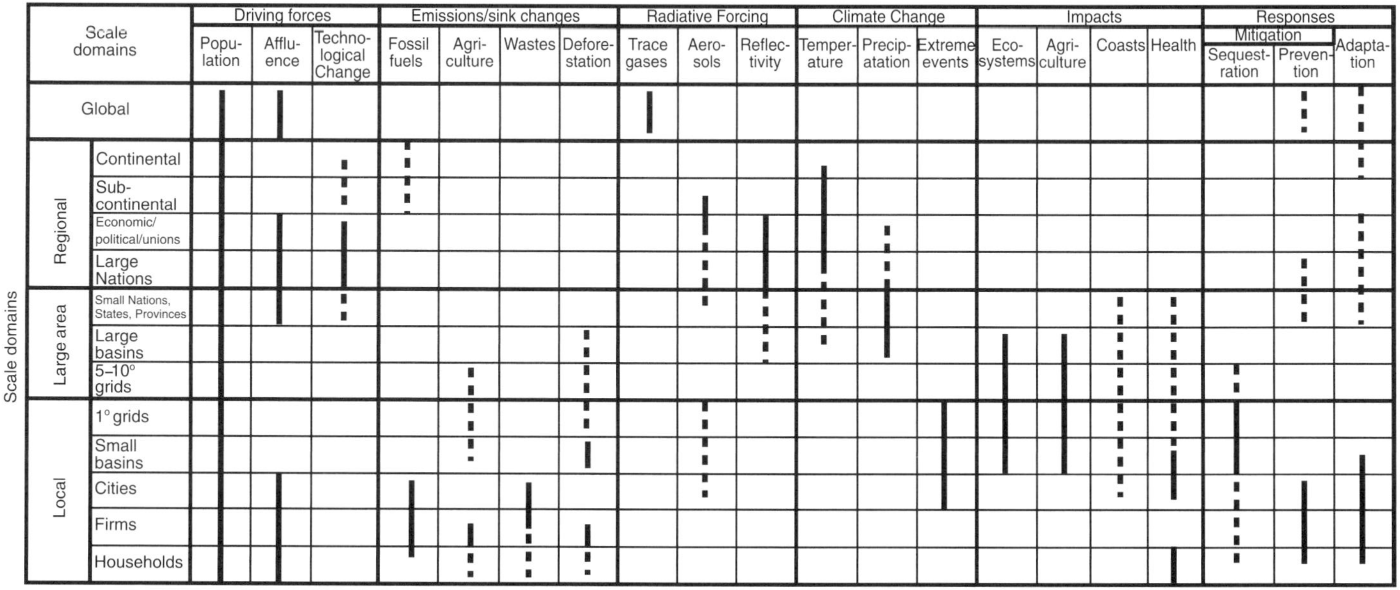

Figure 1.2 Scale domains of climate change and consequences. Depicts the scale of actions, not necessarily the locus of decision-making. Dashed lines indicate occasional consequences or a lower level of confidence.

evolved beneath a greenhouse-like atmosphere. Short-wave solar radiation passes through the atmosphere and is absorbed by the Earth's surface, which it warms. The Earth then radiates energy to space as long-wave infrared radiation. Minor gases in the atmosphere (water vapor, carbon dioxide, ozone, methane, and nitrous oxide) that are transparent to incoming solar radiation absorb and re-emit some of the outgoing long-wave radiation to again warm the Earth's surface and its lower atmosphere. This natural greenhouse effect warms the Earth by as much as 33 °C (91 °F), thus making much of life on Earth possible. Human actions have altered and are continuing to alter natural biogeochemical cycles in ways that increase the concentration of trace gases in the atmosphere.

Viewing the Earth's greenhouse effect and its consequences as a causal chain (Figure 1.2) highlights six major links:

- the societal *driving forces* of
- *human activities* that produce greenhouse gas emissions. Greenhouse gas emissions then provide
- the enhanced *radiative forcing* that
- induces *climate change*, which
- *impacts* nature and society. Finally, the anticipation and experience of the effects of climate change encourage
- a range of *human responses* to prevent climate change, mitigate it, or adapt to it.[8]

Since the onset of the industrial revolution, *human activities* that generate greenhouse gases have increased the concentration of those gases in the atmosphere. According to the third assessment report of the Intergovernmental Panel on Climate Change (Houghton *et al.* 2001) methane has increased by 145%, carbon dioxide by 31%, and nitrous oxide by 16%. In addition, new gases not found in nature (primarily halocarbons, that is carbon compounds containing bromine, chlorine, fluorine, or iodine) have been released into the atmosphere. Current estimates of enhanced *radiative forcing*[9] from these greenhouse gases since pre-industrial times are $+2.425$ W m^{-2}. Carbon dioxide accounts for 60% of this enhanced forcing, methane for 20%, nitrous oxide for 6%, and halocarbons for 14%. Fossil fuel extraction and combustion releases the largest quantity of greenhouse gases, followed in order of importance by deforestation and other land cover changes, agriculture (including cattle rearing, rice production, and fertilizer production and use), industrial production, waste disposal, and refrigeration and air conditioning. Particulates from fossil

---

[8] This chain is shown as a linear sequence for simplification. It consists, of course, of a complex set of relationships with feedbacks at every link affecting other links. More elaborate and operative models of all or most of these links are found in integrated assessments that are designed to inform various stakeholders of alternative courses of human action related to climate change and their associated costs and benefits (Parson and Fisher-Vanden 1997). At least 15 major integrated assessments are underway worldwide. While they differ markedly, systematic comparisons of their inputs and outputs have begun (Toth 1995). A common characteristic of these models is their large scale: two thirds of them, in fact, are global- or continental-scale models (Morgan and Dowlatabadi 1996).

[9] Radiative forcing is a measure of the effect (in watts per square meter) a factor has in altering the balance of incoming and outgoing energy in the Earth–atmosphere system. A positive forcing warms the surface and a negative forcing leads to cooling.

fuel combustion and biomass burning, particularly sulfate aerosols, reflect incoming solar radiation and lead to cooling that offsets positive radiative forcing. Ozone depletion in the stratosphere caused by halocarbons also reduces radiation forcing, while ozone increases in the lower atmosphere (troposphere), mainly from fossil fuels, augment forcing.

The human activities responsible for these greenhouse gas changes are driven by *forces* that have been widely generalized as the **I = PAT** identity (Ehrlich and Holdren 1972), in which changes in **I**mpacts (emissions in this case) are a function of **P**opulation, **A**ffluence, and **T**echnology that increases or decreases impacts or emissions per capita.[10] These *driving forces*, population, affluence, and technology, are again intermediate to more basic driving forces: the complex array of interdependent cultural, economic, environmental, and social contexts examined in Chapter 9 of this volume. Globally, a study using country data for 1989 as observations found average emissions to be driven almost equally by population and affluence (measured by per capita gross domestic product). The effect of population increases with size, however, while the effect of affluence decreases in the richest countries (Dietz and Rosa 1997).

According to the Third Assessment Report (Houghton *et al.* 2001), the indicators of *climate change* over the past century include:

- an increase in global mean surface temperature of 0.4–0.8 °C (0.7–1.4 °F) since about 1860;
- the twentieth century is likely to have been the warmest century in a thousand years in the Northern Hemisphere, with the 1990s the warmest decade, and 1998 the warmest year;
- nighttime temperatures have increased twice as fast as daytime temperatures and the frost-free season has increased in many mid- and high-latitude regions;
- decreases in Northern Hemisphere snow cover (10% since the late 1960s), Arctic Sea ice (10–15% since the 1950s), and alpine glaciers (almost everywhere), but no trends evident in Antarctic sea ice;
- sea level has risen 10–20 cm (4–8 in) since 1900, and this rate of increase is about ten times larger than the average rate of the past 3000 years;
- precipitation has increased by 0.5–1.0% per decade in the twentieth century over most of the mid- and high-latitude Northern Hemisphere, with an increase in heavy and extreme precipitation events and possible flooding, but no increase is evident in hurricanes or severe cyclonic storms;
- warm *El Niño* events have been more frequent, intense, and persistent since the 1970s; and
- all of these climate changes have already impacted avian, insect, plant, and animal life in aquatic, terrestrial, and marine environments on all continents.

---

[10] While there has been widespread use of the I = PAT identity at a global scale, when one moves down scale even to large regions, the complexity and richness of explanation increases. For example, in a study of nine environmental zones under great environmental stress (Kasperson *et al.* 1995), the range of explanatory variables expands beyond population, affluence and technology to include the economic, social, and political institutions that govern resource and environmental use, along with belief systems and attitudes. Poverty emerges as the obverse of affluence and a major driving force in its own right.

These and other indicators, particularly the similarity between observed climate changes and simulated climate from model runs over the past thousand years that included human-induced driving forces as well as natural forces (solar output and volcanic eruptions), led the Intergovernmental Panel on Climate Change's Third Assessment to conclude that 'increasing concentrations of anthropogenic greenhouse gases have contributed substantially to the observed warming over the last fifty years.' Moreover, global average temperature will rise from 1990 to 2100 by about 1.5–6.0 °C (2.7–10.8 °F), and a sea level rise of 14–80 cm (5.5–31.5 in) is possible under a range of emission scenarios. Precipitation will continue to increase, especially over northern mid- and high latitudes, and will increase in some low-latitude regions and decrease in others. Many types of extreme events will also increase, as will continued melting of snow and ice. The potential for large-scale and abrupt changes has been identified through such mechanisms as slowing of the ocean circulation that transports warm water to the North Atlantic, disintegration of the west Antarctic ice sheet, and releases of terrestrial carbon or methylhydrates from permafrost regions or coastal sediments (McCarthy et al. 2001).

Projecting these and similar changes into the future, the Intergovernmental Panel on Climate Change's Third Assessment finds many ways in which natural and human systems become more vulnerable, including:

- as climate changes, ecosystems or biomes are not likely to move as a whole; instead species composition and dominance will change, yielding ecosystems markedly different from ones seen today. In particular, endangered species, arid and semi-arid areas, wetlands overlying permafrost, boreal forests, and coastal and marine ecosystems – especially coral reefs, salt marshes and mangrove forests – appear most vulnerable;
- sea-level rise will increase the vulnerability of some coastal populations to flooding and erosional land loss, especially in deltaic regions and small island states;
- climate change may adversely affect human health due to heat waves and increases in the transmission of such vector-borne infectious diseases as malaria, dengue fever, leishmaniansis, mosquito-borne encephalitis, and cholera and diarrheal disease, because of increases in the active ranges and seasons of their vectors;
- on the whole, global agricultural production could be maintained in the face of climate change but not in the subtropical and tropical areas that are home to many of the world's poorest people. There, changes in climate extremes can lead to major increases in vulnerability through increases in such natural hazards as heat waves, drought, flooding, storm surges, coastal erosion, and possibly cyclones, while in a few regions, vulnerability will be moderated by decreases in cold waves and frost days (McCarthy et al. 2001).

As climate changes or is expected to change, many human-initiated *responses* may follow that are intended to prevent or mitigate the consequences of climate change, or to adapt to climate changes that cannot be prevented or mitigated. Mitigation efforts will focus on human intervention to reduce emissions or on enhancing the action of greenhouse gas sinks, usually by augmenting carbon uptake in forests, soil, and perhaps the oceans. Over the next century the world's energy systems will be replaced twice over, and major reductions in emissions could come from shifting from high-carbon coal to low-carbon oil, lower-carbon gas, and no-carbon nuclear power, or to no-net-carbon wood and plant materials and the no-carbon renewable sources: sun, wind, and water. The efficiency of converting fossil fuel to electricity could double to 60% from its current 30%. Ten to twenty percent of carbon emissions cumulating between now and 2050 could be prevented or stored in such land covers as forests, rangelands, and crop lands. In the short run, combinations of technologies have the potential of reducing global greenhouse gas emissions close to or even below those of the year 2000 by 2010 and even lower by 2020. In the long run, a combination of known technological options and needed socioeconomic and institutional changes can achieve stabilization of the Earth's atmosphere in the range of 450–550 parts per million (ppm) of carbon by volume, a doubling of pre-industrial greenhouse gases. Even if this is achievable, significant adaptation will be required to cope with the impacts already observed and with the many additional changes in physical, biological and human systems that will take place over this century.

Scale in climate change: action and assessment

Worldwide climate change is but one of many environmental changes that are part of the remarkable global changes underway in population, health and well-being, urbanization, economies, technologies, cultures, politics, and institutions (National Research Council 1999). Two major pathways transform regional environmental problems into global problems (Turner *et al*. 1990). *Cumulative* global environmental change begins with common but widespread local problems such as groundwater depletion, pollution, or species extinction. When localized change accumulates to a significant fraction of the total global area or resource, cumulative global change results. *Systematic* global changes are direct alterations in the functioning of a global system exemplified by the effects of greenhouse gas emissions on global climates or the ways ozone-depleting gases affect the stratosphere. Addressing the major systemic changes in global climates and their causes and consequences within domains of nature and society, Clark (1985, 1987) characterized the domains of climate, ecology, and society in terms of their geographic and temporal scales of operation (Figure 1.3). The basic mismatch of scales in space and time at which essential elements of global climate change operate is striking (Wilbanks 2002).

Using the Global Change and Local Places causal structure as a template (Figure 1.2), scale in climate change and its consequences vary more than a billion-fold, from the areal extent of a household, farm, or factory, to the Earth as a whole. Each of the causal links in the chain of driving forces and human responses operates at a characteristic scale within which most actions inside that domain occur. As currently practiced, climate change assessment also uses characteristic scales for each domain, and the relationships between the scales of observation, research, and policy, and the scales of major activities in each link in the causal chain, are critical. The scale at which most action related to each of the domains takes place may well differ from the scale at which decision-making for these actions occurs. As noted above, the scale of action is often smaller than that of the structural context within which action takes place (Figure 1.2).

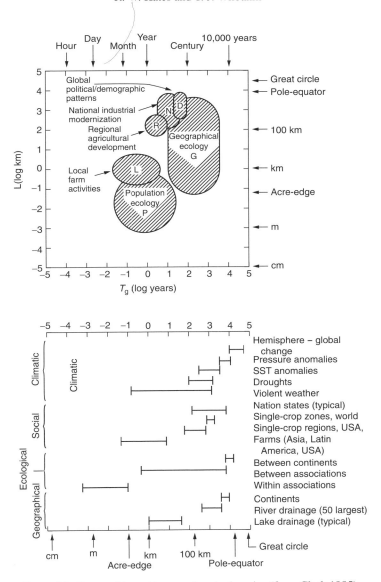

Figure 1.3 Geographic and temporal scale domains (from Clark 1985).

The Global Change and Local Places project examined the scales of action and assessment at four levels: global; regional (continental, subcontinental, economic and political unions, and large nations); large area (small nations, states, provinces, large river basins, and the 5–10° grids commonly used in global climate models); and local (1° grid squares, small river basins, cities, households, farms, firms, and factories). Considerable overlap occurs in such a qualitative scale classification, of course, but it does broadly distinguish differences in scale by size and by common geographic and social units.

Driving forces

The driving forces of population, affluence, and technology (Chapter 9) are intermediate, driven in turn by interdependent cultural, economic, environmental, and social imperatives. Population serves as a driving force because people and households require energy and materials to subsist, some of which release greenhouse gases when produced and used. While the amounts differ greatly among societies, each additional person or household requires some increment of resources and emits a modest amount of carbon dioxide. Affluence drives climate change insofar as it expands demand for energy and materials that release greenhouse gases. Technology is a driving force insofar as different energy and materials production and consumption technologies release markedly different kinds and quantities of greenhouse gases.

Almost all population-creating activity is highly localized in households or their equivalents. Much of the production and consumption enabled by affluence takes place in households, farms, and factories, whereas the technologies most related to climate change operate over larger (but far from universal) areas. Even such global features as automobiles or electricity generation embrace an enormous range of energy and emission efficiencies over large areas and small regions. Within Europe, for example, France emits only half the per capita carbon dioxide that Germany does because of its widespread use of nuclear power. Automobiles in the United States sold in 1995 averaged 8.6 km l^{-1} (20.4 mpg) (United States Environmental Protection Agency 2002), compared with 13.6 km l^{-1} (31.9 mpg) in Europe in the same year (Perkins 1998).

Population numbers and growth are enumerated locally and aggregated to larger areas, from birth registries in industrialized countries, survey data in developing countries, or by censuses everywhere. The populations of almost all localities are known within 20% and in countries with modern statistical services within 3% – better estimates than exist for any other living things and for most other environmental concerns. As a driving force, population can be projected reliably over the short and long term as well as or better than any other aspect of human behavior. Much is also known about the number of children people want and their reasons for having or not having them. Detailed health and demographic surveys of representative national samples have been taken in 27 developing countries. In the policy realm, clear and reliable prescriptions for slowing population growth have been formulated.

Affluence (gross domestic product in the aggregate or disposable income for households) is relatively well measured, but the links between income and the demand for energy and materials that emit greenhouse gases vary greatly among societies and technologies. Much is known at global scale about energy transformations, due in part to common units for conversion among different technologies (Nakicenovic *et al.* 1998). Detailed estimates of energy use are available for countries, regionally, and for the entire world. The forces that drive energy use have been decomposed for industrialized countries and those parsings also highlight the substantial differences in available energy technologies among places (Schipper and Meyers 1992; Schipper *et al.* 1997).

For materials, aggregate data in common units do not exist on a global basis, except for some specific items including materials for energy production, construction, industrial minerals and metals, agricultural crops, and water (World Resources Institute *et al.* 1998). Calculations of material use by volume, mass, or value yield different trends. For limited classes of materials (for example, forest products in the United States), studies of major changes in technological efficiency over time are available (Wernick *et al.* 1997). Overall, the driving force of population is best observed and understood at all scales, whereas the relationship between income and the demand for goods produced by energy and material technologies that emit greenhouse gases is only partly understood, and is generally observed only for large areas.

Emissions and land cover change

To meet population needs and the demands of affluence, people undertake production and consumption that emit greenhouse gases. The greenhouse gases carbon dioxide, methane, and nitrous oxide are released mainly in fossil fuel production and use (manufacturing, electricity generation, transportation, and household heating), forestry and agriculture (land clearing, timber production, wetlands, livestock raising, and fertilizer application), and waste disposal (landfills and incineration). In addition, ozone-depleting chemicals are manufactured and used in a variety of industries, household appliances, and vehicles. Much fossil fuel combustion also emits airborne particles (mainly sulfate aerosols) that act regionally to counter greenhouse warming. All this takes place at the most local of scales: in power plants, factories, vehicles, buildings, households, fields, forests, and animals that constitute billions of point or small area sources of emissions, aerosols, and instances of land cover change.

Analysts usually estimate greenhouse gas emissions by tracking a process that emits the gases, converting the process measure to greenhouse gas releases, and then normalizing the different gases to greenhouse warming potential, or carbon dioxide equivalents. A well-established procedure for such estimates has been developed under the aegis of the Intergovernmental Panel on Climate Change (1992) to meet national reporting requirements of the Framework Convention on Climate Change. Thus estimates of carbon dioxide emissions from fossil fuel consumption and cement production are now available for all countries. The estimates have been extended to smaller areas in some countries, notably by the United States Environmental Protection Agency (1995), in order to calculate state emissions, which now are available for 35 of the 50 United States. Carbon dioxide emissions estimates, though made at local scales, are not true estimates of emissions. The two available sets of $1° \times 1°$ carbon dioxide emissions data for the entire world are national estimates allocated to each $1°$ grid cell in proportion to the estimated share of total population residing within that cell (Andres *et al.* 1996; Olivier *et al.* 1997). Large area and many point source aerosol estimates, particularly for sulfates, are available for most industrialized countries (Graedel *et al.* 1995). Land cover data are more localized, some at the scale of a square kilometer based on satellite observations (http://atlas.esrin.esa.it:8000/) or at even finer scales such as 30 m resolution imagery for the coast of the United States (Dobson *et al.* 1995).

At the global level, past trends in carbon dioxide concentrations in the atmosphere have been estimated for hundreds of millennia, carbon dioxide emissions have been estimated for the 250 years since 1750, regional aerosol data have been calculated in industrialized countries for half a century, and satellite estimates have been made of land cover change over most of the world for the period since 1980. Forecasts of future emissions are available in Intergovernmental Panel on Climate Change reference scenarios for the globe and for major regions based largely on the I = PAT driving force variables: population, economic growth, and technological change. Regional air pollution models for North America, for Europe, and most recently for Asia, provide similar data for future aerosol distributions (McDonald 1999; Tuinstra *et al.* 1999). A major research project aspires to build similar models for land cover change (Turner *et al.* 1995).

Overall, a gross mismatch exists between the billions of point and small area sources of emissions on the one hand, and on the other hand the aggregated data on greenhouse gas and aerosol emissions for nations, regions, and the world, and the assessments and policy analyses that have been based upon those coarse data. Only land cover data are effectively localized, but little is understood about land cover change resulting from deforestation, agriculture, grazing, and urbanization and the resulting emissions of greenhouse gases.

Radiative forcing

While such trace gases as carbon dioxide, methane, nitrous oxide, and ozone-depleting chemicals originate from local sources and are estimated at national and regional scales, they diffuse rapidly in the atmosphere. Consequently, they can be measured globally. Carbon dioxide has been observed in the atmosphere since 1958 and other trace gases (methane, nitrous oxide, and ozone-depleting chemicals) since 1978. Sulfate aerosols arise over large areas, are concentrated in urban and industrial regions, and are transported regionally; they therefore act regionally to counter greenhouse warming potential. Greenhouse gases generated by local changes in land use also diffuse into the atmosphere rapidly, and changes in the albedo (reflectivity) of the earth's surface must be extensive over large areas to significantly affect global climate. The extent of Arctic and Antarctic sea ice may affect albedo.

Much remains to be learned about the distribution of some gases and aerosols in the atmosphere, but current observations can be fed into climate models to estimate the enhanced radiative forcing of the climate system that results from human-induced emissions. Overall, the fit is good between the scale of radiative forcing in the atmosphere, what is observed and what needs to be known for scientific understanding of the atmosphere, and for policy formulation and evaluation with respect to the atmosphere.

Climate change

The three features of greatest interest in climate change are temperature, precipitation, and extreme weather events. For each of the three, many characteristics of interest exist: temperature changes in the stratosphere, troposphere, land surface, and oceans; in land and sea; in the Northern and Southern Hemispheres; at low and high latitudes; between day and night; and between winter and summer, among others. Each of these dimensions changes in response to radiative forcing in different directions or at different rates and scales, and the patterns of such changes might be the best confirmation of human-induced climate change.

Clark (1985) sought to compare some of the spatial domains of climate events and found (Figure 1.3) that temperature change as evidenced by historic warming trends appears in areas greater than 10,000 square kilometers (3,850 square miles), precipitation or its absence (as in major droughts) takes place in areas of 1,000 – 10,000 square kilometers (385 – 3,850 square miles), and extreme weather occurs at a scale of 0.1 – 1,000 square kilometers (0.04 – 385 square miles). Overall, the characteristic scales at which climate change occurs vary a million-fold.

The match between the active domains of climate events and the scales at which major parameters of climate are observed, aggregated, and modeled, also varies greatly. The best fit exists for currently observed climate events, given a dense web of observing stations on land, an increasing number at sea, and the availability of satellite observations since 1979.[11] These data can be aggregated into relatively homogenous large areas such as the 344 climatic divisions of the mainland United States, which match well the scales of temperature and precipitation change. The 140 year instrument record has now been placed in a context of millennial length by such proxy variables as tree rings and corals, and in a frame extending back hundreds of thousands of years by gases trapped in ice caps (Crowley 2000).

But crucial projections of future climates caused by radiative forcing from enhanced emissions depend solely on complex models of the Earth's climate. The current resolution of these models (5° grids at best) are thought to be reliable primarily for temperature, and only over large latitudinal bands or continental zones. Overall, climate itself is measured as well as or better than any of the causal facets of climate change, and those measures can be aggregated to match the scale of climate events. Yet there remains a large gap between climate change forecasting models and the scales at which extreme weather and long-term climate are experienced.

Impacts

Changes in temperature and precipitation and extreme weather profoundly affect natural and managed ecosystems and human activities and well-being. The Intergovernmental Panel on Climate Change assessment (McCarthy *et al.* 2001) focused on the vulnerability of seven natural and human systems that include the major terrestrial and marine ecosystems and that provide water, food and fiber, human infrastructure, and health, as well as on eight continental and larger-sized regions. Climate helps define the areal extent of many of these impacted systems, and effects of climate change impacts may appear first as changes at the margins of ecosystems, crop regions, shorelines, or disease vector habitats. The scale

---

[11] Many areas of the world remain only sparsely monitored, many records need to be revised for reliability and to remove site changes and urban effects, and much needs to be learned about relationships between satellite and ground observations (National Research Council, Panel on Reconciling Temperature Observations 2000).

of these climate-bounded areas varies considerably, but tends to approximate large areas or their borders. Large ecosystems in the United States, for example, may vary within the same size range as the 50 states. In one classification, the 52 ecoregions of the United States range from the 9,600 square kilometers (3,700 square miles) of the Black Hills of South Dakota, to the 751,000 square kilometers (290,000 square miles) of the Great Plains – Palouse Dry Steppe region (Bailey 1995). Major agricultural crop regions approximate the largest of the 52 ecoregions, encompassing three to eight states in areal extent (United States Department of Agriculture 1987). Coastal zones are very different sized regions. The narrow but continuous slivers of land subject to a 1 m rise in sea level range from several square kilometers on small islands, through thousands of square kilometers for medium-sized countries, to tens of thousands of square kilometers in low-lying areas in Bangladesh, China, and the United States. Climate changes will also alter the habitats of insect and animal vectors of human and biotic diseases. Such changes might take the form of increased bands of malarial infestation or narrowed zones of river-constrained onchocerciasis. Similarly, changed or intensified tracks of such extreme weather events as hurricanes, tornadoes, hail, and wind would also vary from localities to regions.

Most climate change impact analyses to date begin with outputs from global climate models or with a hypothesized arbitrary change in temperature, precipitation, or sea level. Then, using models or analogs, impact analysts try to assess positive and negative impacts on natural ecosystems and human activities. Because of the coarse resolution and poor reliability of climate models for large areas, impact analysis (especially Intergovernmental Panel on Climate Change assessments) have focused on major ecosystem types, generic economic sectors, or very large regions – usually at continental or subcontinental scales. Agricultural impact studies have been made for major crop types and ecosystems, and for large biomes. A notable exception has been studies of sea level rise, which can be highly locality-specific, though sometimes complicated by local uplift and subsidence. In general, impacts that occur at local or regional scales are poorly matched by generic assessments, which are rarely locality-specific.

Responses

People and societies will in time come to anticipate climate changes and their consequences and will seek to prevent change, mitigate its extent, or adapt to and reduce their vulnerability to such changes. Preventive actions to remove greenhouse gases from the atmosphere or to change the Earth's radiation balance are often called *geoengineering*. Actions to remove carbon by collecting it in the course of energy production and sequestering it in the ground or the deep oceans have shown substantial progress and at least one current application in Norway (Parson and Keith 1998). The most feasible immediate prevention strategy appears to be the removal of carbon from the atmosphere by increasing carbon storage in forests, soil, and perhaps the oceans. To be effective in preventing or ameliorating climate change, storage must be enhanced over large areas. Yet the actions that must be taken to accomplish enhanced storage – tree planting, reversion of fields to forests, and stimulation of photosynthesis – must be accomplished locally even if they eventually cumulate to large areas.

Reducing greenhouse gas emissions, a major focus of mitigation, also requires action at the billions of point sources where emissions occur. Creating structures that encourage such local actions requires international agreements, national policies, corporate decisions, and public support: efforts that span all scales from local to global. Adaptation, even more than abatement, takes place locally, but similarly can be encouraged by global, national, and corporate policies and by public attitudes. Human responses to climate change will depend on a combination of local decision and actions and state, national, and global mandates and enabling policies.

To date, most mitigation studies have been conducted for rather large areas. Research on carbon storage or enrichment in forests or fields is highly localized, but studies of storage potential are usually based on large-area ecosystems and land uses. Greenhouse gas abatement studies are usually generic rather than place-specific, organized by technological or economic sectors. Thus volume three of the third Intergovernmental Panel on Climate Change report (Davidson *et al.* 2001) has chapters that address mitigation generically in buildings, transport, industry, agriculture, waste management, and energy supply. Similarly, the second volume of the Intergovernmental Panel on Climate Change report (McCarthy *et al.* 2001), which addresses impacts, adaptation, and vulnerability, considers adaptation for each of the seven systems and eight regions it addresses, but with rare exceptions, only in the most generic ways. Thus the mismatch between the highly localized scale of human responses on the one hand, and generic assessments of the range of human mitigation and adaptation options on the other, remains serious.

To sum across the links in Figure 1.2, an envelope of the larger-scale actions (global and regional) appears as a wave in which global and large regional actions characterize the driving forces of population, affluence, and technological change, the radiative forcing of gases, aerosols, and reflectivity, climate change, and preventive and adaptive responses. In contrast, emissions and sinks and the major impacts of climate change are far more localized.

Addressing the mismatch in scale domains

Thus for emissions impacts and most responses, there is a grave mismatch between the scale domains of human activity on the one hand, and of observation, research, and policy assessment and formulation on the other. This gap between the knowledge that is needed to act locally and what is currently being done globally to generate knowledge about climate change and its impacts is increasingly recognized as an impediment to further progress. Efforts are therefore underway to move *down* scale in each of the causal domains.

Downscaling top-down approaches

Driving forces

Decomposition analysis identifies the differing importance of the various driving forces of greenhouse gas emissions. It has been applied to the forces driving eighteen years of carbon dioxide production in ten countries, and has been used to identify major differences in sources and in emissions even among highly industrialized countries (Schipper and Myers 1992; Schipper *et al*. 1997). At a larger regional scale than countries, six driving forces (population growth, economic growth, energy intensity, technological change, resource base, and environment) are used as determinants of future energy systems for 11 world regions (Nakicenovic *et al*. 1998).

Emissions and land cover change

Efforts are underway to complete the estimation of greenhouse gas emissions using standardized Intergovernmental Panel on Climate Change methodology for all nations in order to downscale these estimates to regions and large areas (Graedel *et al*. 1993, 1995). Estimates are now available for all countries based on population figures when direct data are lacking (Carbon Dioxide Information and Analysis Center 2002). In 56 developing and transitional countries, studies are underway intended to create national capacities for assessment and to identify selected regional or sectoral impacts of climate change, in addition to estimating greenhouse gas emissions (Dixon *et al*. 1996; United States Country Studies Program 2002). In the United States, the Environmental Protection Agency has downscaled the Intergovernmental Panel on Climate Change emissions methodology, making appropriate modifications for calculating state (large-area) emissions, which have been completed or are underway in 35 states (United States Environmental Protection Agency 1999). Similar steps have been taken in Australia (Australian Greenhouse Office undated). Gradually, downscaling has moved farther down, particularly to city and metropolitan areas (McEvoy *et al*. 1997). There is increasing interest and action on the part of corporations to inventory their own emissions, to register or publicly disseminate changes in their emissions, and in some cases to create intercorporate trading in emissions regimes (Loreti *et al*. 2000). A quarter of the increase in carbon dioxide emissions over the past 20 years is attributed to land use change, and satellite imagery has provided continuous coverage of the earth since 1992 with the potential to measure changes in land cover and use for areas as small as one square kilometer, or 0.39 square miles (Global 1KM AVHRR Server 2002).

Climate change

A major effort is underway to downscale global climate change models to yield more credible forecasts of large-area climate changes and impacts via two major approaches. The *nested* approach employs large-area regional models (Girogi *et al*. 1994; Jenkins and Barron 1996) that simulate regional topography, vegetation, and water bodies nested within larger global models. Alternatively, *statistical* relationships can be used to link the major features of global models with more local aspects of climate or weather. The recent United States national assessment of climate change used two different models to downscale climate changes for 19 different regions of the country (National Assessment Synthesis Team 2000). Some models in the United States currently forecast at a scale of a 50 km grid, while experiments are being undertaken in Japan with a 10 km grid.

More global change researchers are now focusing on short-term climate forecasting, moving from the decades-to-centuries perspective of greenhouse warming to seasonal and interannual forecasts, whose reliability for certain areas has improved considerably. Recent El Niño forecasts anticipated the 1997–8 ocean warming by as much as six months, although they performed less well later into the event (Kerr 2000). Such forecasts are more relevant locally and regionally, and when reliable, build user confidence in undertaking anticipatory responses and adaptation to climate change.

Impacts

Efforts to identify place-specific impacts of global climate change are increasing (Rosenzweig and Hillel 1998). Using climate model results to create a regional or large area scenario has traditionally combined model outputs that vary widely from model to model and provide a range of potential regional climate changes on which to base impact assessments. More promising are the major efforts underway to downscale climate model outputs noted above. Another option is to use analogs from the past (historic climate events) or from other places (climate-bounded ecosystems or economies) to simulate regional impacts of climate change. In the MINK (**M**issouri, **I**owa, **N**ebraska, and **K**ansas) study, the weather during the great drought of the 1930s was applied to the current ecosystems, economy, and population of four states (Rosenberg 1993). A study of the Mackenzie River basin in Canada used a recent period of warming and its observed impacts to simulate long-term impacts of global warming in the region (Cohen 1997). The Holdridge triangle of life zones has often been used to forecast place-specific changes in ecosystems based on anticipated changes in temperature and precipitation (Holdridge 1947, 1967; Emanuel *et al.* 1985; Pitelka 1997). Increasingly, efforts to assess impacts have begun to focus on regions smaller than the continental size areas of the Intergovernmental Panel on Climate Change Third Assessment. In the United States, the recently concluded National Assessment of Climate Change impacts began with 19 regions that were later aggregated into nine regions (including the scattered holdings of native Americans) and five activity sectors (National Assessment Synthesis Team 2000). In Canada, impacts were assessed for six regions and 12 sectors (Maxwell *et al.* 1997). There is a growing library of national impact studies around the world.

Responses

Although attention still focuses on the human responses required by such international agreements as the Framework Convention on Climate Change and the Kyoto Protocol, downscaling is evident in addressing responses, in this case making international agreement more difficult to attain. Blocs of nations (the so-called *umbrella group* led by the United States, the European Union, the G-7 group of developing countries, the small islands, and the oil producers) each advocate quite different response strategies, which has led to the collapse of talks to implement the Kyoto Protocol to reduce greenhouse gases worldwide by 5% below 1990 levels by 2010–12. Within countries, action plans are increasingly based on regions. In the United States, for example, some 19 regions (National Assessment Synthesis Team 2000) and 35 states have begun to formulate responses appropriate for their territories. In Canada, a study of adaptation considers six regions of the country (Maxwell *et al.* 1997). Such geographical downscaling is paralleled by sectoral downscaling. Climate impacts and adaptation in agriculture vary from place to place, and theoretically these impacts and adaptations are assessed along with other factors in the price of land. Using these differences, a set of studies have estimated impacts and adaptations for areas as small as counties in the United States (Mendelsohn *et al.* 1994, 1999; Polsky and Easterling 2000).

Upscaling bottom-up approaches

In recognition of the mismatches in scale among important domains of climate change, its consequences, and human responses, a growing interest in creating bottom-up approaches is evident. These have included local governmental, non-governmental, and corporate efforts, as well as the project that this volume summarizes.

Cities for climate protection

Local governments vary considerably both in their competence to undertake greenhouse gas reduction and in their willingness to do so (Collier and Löfstedt 1997). The most extensive current effort if this kind on the part of local governments is part of the international policy initiative entitled The Cities for Climate Protection Campaign, which has fostered an asphalt-roots movement (Chapter 12). The Cities for Climate Protection Campaign originated with the International Council for Local Environmental Initiatives Urban Carbon Dioxide Reduction Project, in which twelve North American and European cities worked together to develop and test methods whereby local governments could implement greenhouse gas emission reduction strategies (International Council for Local Environmental Initiatives 1996). Based on that project's results, the International Council for Local Environmental Initiatives established in 1995 the Cities for Climate Protection Campaign to bring together local governments committed to greenhouse gas emission reductions. By late 1999, the rapidly growing program had 403 members worldwide, including 68 in the United States. Together, these cities account for an estimated 8% of global carbon dioxide emissions.

Cities, counties, or metropolitan regions join the campaign by formally resolving to complete five key tasks: (1) an energy and emissions inventory; (2) a forecast of future emissions; (3) adoption of emissions reduction targets; (4) plans for local actions to achieve the reduction targets; and (5) implementation of those actions to reduce carbon dioxide and methane. To support such local government efforts, The Cities for Climate Protection Campaign has created analytical tools that allow municipalities to track their own emissions, forecast changes over time, and assess the potential impacts of diverse technical and policy measures designed to meet their target reductions.

Equally important, and perhaps of even greater importance in the long run, are corporate efforts to reduce greenhouse gases led by oil companies such as BPAmoco, Shell, and

Sunoco; energy companies such as Enron and American Electric Power; and such industrial giants as Boeing, Dupont, IBM, and Toyota. These private sector programs are facilitated by such groups as the Business Environmental Leadership Council of the Pew Center on Global Climate Change, and the World Business Council for Sustainable Development – World Resources Institute greenhouse gas protocol effort (http://www.ghgprotocol.org). Companies such as BPAmoco have adopted greenhouse gas emissions reduction targets of 10% below 1990 baseline emissions by 2010 and have created internal and external trading regimes to achieve these goals.

Non-governmental efforts are widespread, symbolized by the laborious effort to create a sandbagged dike across from the meeting hall at The Hague in The Netherlands at the crucial conference to implement the Kyoto Protocol, to remind delegates of the reality of global warming and sea-level rise. Non-governmental efforts are not restricted to national and international lobbying, but also create opportunities for individuals and their households to take steps to reduce their greenhouse gas emissions, partly by developing tools to calculate personal and household emissions.

Upscale and downscale approaches come closest together in the work emerging on the concept of vulnerability. These efforts seek to characterize vulnerability: the susceptibility to injury, damage, or harm of ecosystems, places, people, livelihoods, or activities. In vulnerability, three important factors come together: sensitivity to climate, exposure to climate change, and resilience or adaptive capacity (McCarthy *et al.* 2001). Unlike conventional downscaled methods of impact analysis that require some output from global climate models applied to a region or locale, vulnerability analyses can begin with the inherent characteristics of the place, group, or activity, and then assess its inherent sensitivity to climate and its capacity to cope with climate change or to respond to it. Even in the absence of a projected exposure, which is usually obtained from a downscaled model, it is possible to estimate the type and magnitude of exposure that would cause harm given inherent sensitivity and adaptive capacity, and to use these to suggest boundaries for climate change that would prevent excessive harm. Much more needs to be done in developing methods to characterize each of these elements at different scales, as well as to go beyond vulnerability to climate change to include vulnerability to the multiple environmental and social stresses that actually confront the places, peoples, and systems of interest (Clark *et al.* 2000). But this is where much of the frontier research will take us.

Global Change and Local Places

In this evolving context, *Global Change and Local Places* reports on a sustained and systematic effort to address the grand query of scale from a bottom-up perspective. The Global Change and Local Places project asked how, when, and where local knowledge, volition, and opportunity can be employed in addressing the great global challenge of human-induced climate change. The project began with the observation that scale domains differ for different parts of the global climate change causal chain. It postulates that understandings of the processes will differ according to the scale of observation, with greater variance, volatility, and value of local knowledge evident at local scales, and that downscaling and upscaling are likely to contribute different insights. And it postulates that scale interactions are significant in global change processes.

Global Change and Local Places does not answer the grand query; if such queries were easily answered, they would hardly be grand. The volume does provide an example of bottom-up research on global change in four quite different parts of the United States, an example that relates the near-universals of the greenhouse effect to the particulars of local emissions and efforts to mitigate them, and an effort that partly unravels the webs of structure and agency, and macro-processes and micro-behavior, that link the global and the local everywhere on Earth.

REFERENCES

Alexander, J. C., B. Giesen, R. Münch, and N. J. Smelser, eds. 1987. *The Micro-Macro Link*. Berkeley: University of California Press.

Andres, R. J., G. Marland, I. Fung, and E. Matthews. 1996. A 1 Degree × 1 Degree Distribution of Carbon Dioxide Emissions from Fossil Fuel Consumption and Cement Manufacture, 1950–1990. *Global Biogeochemical Cycles*, **10**: 419–29.

Aronson, E., and P. C. Stern. 1984. *Energy Use: The Human Dimension*. San Francisco: W. H. Freeman.

Australian Greenhouse Office, undated. *Supplementary Methodology for State and Territory Inventories Based on Workbook 1.1 Energy Workbook for Fuel Combustion Activities (Stationary Sources)*. Canberra, ACT: Australian Greenhouse Office.

Bailey, R. G. 1995. *Description of the Ecoregions of the United States*. Washington: United States Department of Agriculture. USDA Forest Service Miscellaneous Publication 1391.

Bohle, H. G., T. E. Downing, and M. J. Watts. 1994. Climate Change and Social Vulnerability. *Global Environmental Change – Human and Policy Dimensions*, **4**: 37–48.

Carbon Dioxide Information and Analysis Center, Oak Ridge National Laboratory. 2002. http://cdiac.esd.ornl.gov/home.html.

Clark, W. C. 1985. Scales of Climate Impacts. *Climatic Change*, **7**: 5–27.

Clark, W. C. 1987. Scale Relationships in the Interaction of Climate, Ecosystems, and Societies. In K. C. Land and S. H. Schneider, eds. *Forecasting in the Social and Natural Sciences*: 337–78. Dordrecht: D. Reidel.

Clark, W. C., J. Jäger, R. Corell, R. Kasperson, J. J. McCarthy, D. Cash, S. J. Cohen, P. Desanker, N. M. Dickson, P. Epstein, D. H. Guston, J. M. Hall, C. Jaeger, A. Janetos, N. Leary, M. A. Levy, A. Luers, M. MacCracken, J. Melillo, R. Moss, J. M. Nigg, M. L. Parry, E. A. Parson, J. C. Ribot, H.-J. Schellnhuber, G. A. Seielstad, E. Shea, C. Vogel, and T. J. Wilbanks. 2000. *Assessing Vulnerability to Global Environmental Risks*. Report of the Workshop on Vulnerability to Global Environmental Change: Challenges for Research, Assessment and Decision Making. May 22–25, 2000, Airlie House, Warrenton, Virginia. Cambridge, MA: Belfer Center for Science and International Affairs (BCSIA) Discussion Paper 2000-12, Environment and Natural Resources Program, Kennedy School of Government, Harvard University, Available at http://www.ksg.harvard.edu/sust

Cohen, S. J. 1997. What If and So What in Northwest Canada? Paper presented at the Annual Meeting of the American Association for the Advancement of Science, Seattle WA, 14 February 1997.

Collier, U., and R. E. Löfstedt. 1997. Think Globally, Act Locally?: Local Climate Change and Energy Policies in Sweden and the U. K. *Global Environmental Change*, **7**: 25–40.

Cox, K. R. 1997. *Spaces of Globalization: Reasserting the Power of the Local.* New York: Guilford Press.

Crowley, T. J. 2000. Causes of Climate Change Over the Past 1000 Years. *Science*, **289**: 270–7.

Cutter, S. 1993. *Living with Risk.* London: Edward Arnold.

Davidson, O., B. Metz, J. Pan and R. Swart, eds. 2001. *Climate Change 2001: Mitigation.* Cambridge: Cambridge University Press.

Dietz, T., and E. A. Rosa. 1997. Effects of Population and Affluence on CO_2 Emissions. *Proceedings of the National Academy of Sciences,* **94**: 175–9.

Dixon, R. K., J. A. Sathaye, S. P. Meyers, O. R. Masera, A. A. Makarov, S. Toure, W. Makundi, and S. Wiel. 1996. Greenhouse Gas Mitigation Strategies: Preliminary Results from the U. S. Country Studies Program. *Ambio*, **25** (1): 26–32.

Dobson, J. E., E. A. Bright, D. W. Field, R. L. Ferguson, K. D. Haddad, H. Iredale III, J. R. Jensen, V. V. Klemas, R. J. Orth, L. L. Wood, and J. P. Thomas. 1995. *NOAA Coastal Change Analysis Program: Guidance for Regional Implementation.* Washington DC: NOAA Technical Report NMFS 123.

Ehrlich, P. R., and J. P. Holdren. 1972. Review of *The Closing Circle. Environment*, April: 24–39.

Emanuel, W. P., H. H. Shugart, and M. P. Stevenson. 1985. Climate Change and the Broad-Scale Distribution of Terrestrial Ecosystem Complexes. *Climatic Change*, **7**: 29–43.

Fischer G., K. Frohberg, M. L. Parry, and C. Rosenzweig. 1994. Climate Change and World Food Supply, Demand and Trade: Who Benefits, Who Loses? *Global Environmental Change*, **4**: 7–23.

Gallagher, R. and T. Appenzeller. 1999. Beyond Reductionism. *Science*, **284** (5411): 79.

Global 1KM AVHRR Server. 2002. http://atlas.esrin.esa.it:8000/

Graedel, T. E., T. S. Bates, A. F. Bouwman, D. Cunnold, J. Dignon, I. Fung, D. J. Jacob, B. K. Lamb, J. A. Logan, G. Marland, P. Middleton, J. M. Pacyna, M. Placet, and C. Veldt. 1993. A Compilation of Inventories of Emissions to the Atmosphere. *Global Biogeochemical Cycles*, **7**: 1–26.

Graedel, T. E., C. M. Benkovitz, W. C. Keene, D. S. Lee, and G. Marland. 1995. Global Emissions Inventories of Acid-Related Compounds. *Water, Air, and Soil Pollution*, **85**: 25–36.

Hewitt, K. 1997. *Regions of Risk: A Geographical Introduction to Disasters.* London: Longman.

Holdridge, L. R. 1947. Determination of World Plant Formations from Simple Climatic Data. *Science*, **105**, 367–8.

Holdridge, L. R. 1967. *Life Zone Ecology.* San Jose, Costa Rica: Tropical Science Center.

Holling, C. S. 1992. Cross-scale Morphology, Geometry and Dynamics of Ecosystems. *Ecological Monographs*, **62**: 447–502.

Holling, C. S. 1995. What barriers? What bridges? In L. H. Gunderson, C. S. Holling, and S. S. Light, eds. *Barriers and Bridges to the Renewal of Ecosystems and Institutions*: 3–34. New York: Columbia University Press.

Houghton, J. T., Y. Ding, D. J. Griggs, M. Noguer, P. J. van der Linden, D. Xiaosu, and K. Maskell, eds. 2001. *Climate Change 2001: The Scientific Basis.* Cambridge: Cambridge University Press.

Intergovernmental Panel on Climate Change, Organization for Economic Cooperation and Development. 1992. *National Inventories of Net Greenhouse Gas Emissions: IPCC Guidelines for Preparation and Reporting.* Paris: OECD.

International Council for Local Environmental Initiatives. 1996. *Urban Greenhouse Gas Inventories and Emission Reduction Assessment in Canada.* Ottawa: ICLEI and Torrie Smith Associates.

Jenkins, G. S., and E. J. Barron. 1996. Global Climate Model and Coupled Regional Climate Model Simulations Over the Eastern United States. *Global and Planetary Change*, **15**: 3–32.

Karl, T. R., and W. E. Riebsame. 1984. The Identification of 10- to 20-year Climate Fluctuations in the Contiguous United States. *Journal of Climate and Applied Meteorology*, **23**: 950–66.

Karl, T. R., and W. E. Riebsame. 1989. The Impact of Decadal Fluctuations in Mean Precipitation and Temperature on Runoff: A Sensitivity Study over the United States. *Climatic Change*, **15**: 423–47.

Kasperson, J. X., R. E. Kasperson, and B. L. Turner, II., eds. 1995. *Regions at Risk: Comparisons of Threatened Environments*. Tokyo: United Nations University.

Kasperson, R. E., R. W. Kates, and C. Hohenemser, 1985. Hazard Management, in R. W. Kates, C. Hohenemser, and J. X. Kasperson, eds. *Perilous Progress: Managing the Hazards of Technology*: 43–66. Boulder: Westview Press.

Kates, R. W. and V. Haarmann. 1992. Where the Poor Live: Are the Assumptions Correct? *Environment*, **4**: 4–11, 25–8.

Kerr, R. A. 2000. Second Thoughts on Skill of *El Niño* Predictions. *Science*, **290**: 256–7.

Levin, S. A. 1992. The Problem of Pattern and Scale in Ecology. *Ecology*, **73**: 1943–67.

Loreti, C. P., W. F. Wescott, and M. A. Eisenberg. 2000. *An Overview of Greenhouse Gas Emissions Inventory Issues*. Arlington, VA: Pew Center on Global Climate Change.

Maxwell, B., N. Mayer, and R. Street, eds. 1997. *The Canada Country Study: Climate Impacts and Adaptation, National Summary for Policymakers*. Ottawa: Environment Canada.

McCarthy, J., O. Canziani, N. Leary, D. Dokken, and K. White, eds. 2001. *Climate Change 2001: Impacts, Adaptation, and Vulnerability*. Cambridge: Cambridge University Press.

McDonald, A. 1999. Combating Acid Deposition and Climate Change. *Environment*, **41** (3): 4–11, 34–41.

McEvoy, D., D. C. Gibbs, and J. W. S. Longhurst. 1997. Assessing Carbon Flow at Local Scale. *Energy and Environment*, **8** (4): 297–311.

Mendelsohn, R., W. Nordhaus, and D. Shaw. 1994. The Impact of Global Warming on Agriculture: A Ricardian Analysis. *American Economic Review*, **84** (4): 753–71.

Mendelsohn, R., W. Nordhaus, and D. Shaw. 1999. The Impact of Climate Variation on U. S. Agriculture. In R. Mendelsohn and J. E. Neumann, eds. *The Impact of Climate Change on the United States Economy*. Cambridge: Cambridge University Press.

Meyer, W. B., and B. L. Turner II, eds. 1998. *Changes in Land Use and Land Cover: A Global Perspective*. Cambridge: Cambridge University Press.

Meyer, W. B., D. Gregory, B. L. Turner II, and P. McDowell. 1992. The Local–Global Continuum. In Abler, R. F., Marcus, M. G., and Olson, J. M. eds. *Geography's Inner Worlds: Pervasive Themes in Contemporary American Geography*: 255–79. New Brunswick, NJ: Rutgers University Press.

Morgan, M. G., and Dowlatabadi, H. 1996. Learning from Integrated Assessment of Climate Change. *Climate Change*, **34**: 337–68.

Nakicenovic, N., A. Grubler, and A. McDonald. 1998. *Global Energy Perspectives*. Cambridge: Cambridge University Press.

National Assessment Synthesis Team. 2000. *Climate Change Impacts on the United States: The Potential Consequences of Climate Variability and Change*, Washington: U. S. Global Change Research Program.

National Research Council, Board on Sustainable Development. 1999. *Our Common Journey: A Transition Toward Sustainability*. Washington: National Academy Press.

National Research Council, Panel on Reconciling Temperature Observations. 2000. *Reconciling Observations of Global Temperature Change*. Washington: National Academy Press.

Olivier, J. G. J., A. F. Bouwmann, C. W. M. van der Maas, J. J. M. Berdowski, C. Veldt, J. P. J. Bloos, A. J. H. Visschedijk, P. Y. J. Zandveld, and J. L. Haverlag. 1997. Description of EDGAR Version 2.0: A Set of Global Emissions Inventories of Greenhouse Gases and Ozone-Depleting Substances for All Anthropogenic and Most Natural Sources on a per Country Basis and on 1 Degree × 1 Degree Grid. Bilthoven, The Netherlands: National Institute of Public Health and the Environment. RIVM Report 771060 002.

Openshaw, S., and Taylor, P. J. 1979. A Million or So Correlation Coefficients: Three Experiments on the Modifiable Areal Unit Problem. In N. Wrigley, ed. *Statistical Methods in The Spatial Sciences*: 127–44. London: Routledge & Kegan Paul.

Palm, R. I. 1990. *Natural Hazards: An Integrated Framework for Research and Policy*. Baltimore: Johns Hopkins University Press.

Parson, E. A., and K. Fisher-Vanden. 1997. Integrated Assessment Models of Global Change. *Annual Review of Energy and Environment*, **22**: 589–628.

Parson, E. A., and D. W. Keith. 1998. Climate Change: Fossil Fuels without CO_2 Emissions. *Science* **282**: 1053–4.

Perkins, S. 1998. Focus on Transport Emissions Needed if Kyoto's CO_2 Targets Are to Be Met. *Oil and Gas Journal*, 19 January 1998: 35–9.

Pitelka, L. 1997. Plant Migration and Climate Change. *American Scientist*, **85**: 464–73.

Polsky, C., and W. E. Easterling. 2000. Adaptation to Climate Variability and Change in the U. S. Great Plains: A Multi-scale Analysis of Ricardian Climate Sensitivities. *Agriculture, Ecosystems and Environment*, **85** (1–3): 133–44.

Rediscovering Geography Committee, National Research Council. 1997. *Rediscovering Geography: New Relevance For Science and Society*. Washington, DC: National Academy Press.

Root, T. L., and S. H. Schneider. 1995. Ecology and Climate: Research Strategies and Implications. *Science*, **269**: 334–41.

Rosenberg, N. J., ed. 1993. *Towards an Integrated Impact Assessment of Climate Change: The MINK Study*. Dordrecht: Kluwer Academic.

Rosenzweig, C., and D. Hillel. 1998. *Climate Change and the Global Harvest: Potential Impacts of the Greenhouse Effect on Agriculture*. New York: Oxford University Press.

Schipper, L. J., and S. Meyers. 1992. *Energy Efficiency and Human Activity: Past Trends, Future Prospects*. Cambridge: Cambridge University Press.

Schipper, L. J., M. Ting, M. Khrushch, and W. B. Golove. 1997. The Evolution of Carbon Dioxide Emissions from Energy Use in Industrialized Countries: An End-Use Analysis. *Energy Policy*, **25** (7–9): 651–72.

Toth, F. L. 1995. Practice and Progress in Integrated Assessments of Climate Change: A Workshop Overview. *Energy Policy*, **23**: 253–67.

Tuinstra, W., L. Hordijk, and M. Amann. 1999. Using Computer Models in International Negotiations: the Case of Acidification in Europe. *Environment*, **41** (9): 32–42.

Turner, B. L. II, R. E. Kasperson, W. B. Meyer, K. M. Dow, D. Golding, J. X. Kasperson, R. C. Mitchell, and S. J. Ratick. 1990. Two Types of Global Environmental Change: Definitional and Spatial Scale Issues in Their Human Dimensions. *Global Environmental Change*, 1: 14–22.

Turner, B. L. II, G. Hyden, and R. W. Kates, eds. 1993. *Population Growth and Agricultural Change in Africa*. Gainesville: University Press of Florida.

Turner, B. L. II, D. Skole, S. Sanderson, G. Fischer, L. Fresco, and R. Leemans. 1995. *Land Use and Land-Cover Change: Science/Research Plan.* Stockholm: International Geosphere-Biosphere Program. IGBP Report #35/HDP Report #7.

Turner, M. G., W. H. Romme, R. H. Gardner, R. V. O'Neill, and T. K. Kratz. 1993. A Revised Concept of Landscape Equilibrium: Disturbance and Stability on Scaled Landscapes. *Landscape Ecology* **8** (3): 213–27.

United States Country Studies Program. 2002. http://www.gcrio.org/CSP/uscsp.html

United States Department of Agriculture, World Agricultural Outlook Board. 1987. *Major World Crop Areas and Climatic Profiles.* Washington: USDA Agriculture Handbook No. 664.

United States Environmental Protection Agency. 1995. *State Workbook; Methodologies for Estimating Greenhouse Gas Emissions, 2nd Edition.* Washington: U. S. Environmental Protection Agency, Office of Policy, Planning, and Evaluation, State and Local Outreach Program. EPA-230-B-92-002.

United States Environmental Protection Agency. 1999. *Inventory of U. S. Greenhouse Gas Emissions and Sinks: 1990–1997.* EPA 236-R-99-003. http://yosemite.epa.gov/oar/globalwarming.nsf/content/ResourceCenterPublications GHGEmissions.html

United States Environmental Protection Agency, Office of Transportation and Air Quality. 2002. Light-Duty Automotive Technology and Fuel Economy Trends 1975 through 2001. EPA420-R-01-008.http://www.epa.gov/otaq/fetrends.htm

Wernick, I., P. Waggoner, and J. Ausabel. 1997. Searching for the Leverage to Conserve Forests: The Industrial Ecology of Wood Products in the United States. *Journal of Industrial Ecology*, **1** (3): 125–45.

White, G. L. ed. 1974. *Natural Hazards: Local, National, Global.* New York: Oxford University Press.

Wilbanks, T. J. 1984. Scale and the Acceptability of Nuclear Energy. In M. Pasqueletti and K. D. Pijawka, eds. *Nuclear Power: Assessing and Managing Hazardous Technology*: 9–50. Boulder: Westview Press.

Wilbanks, T. J. 1994. 'Sustainable Development' in Geographic Context. *Annals of the Association of American Geographers*, **84**: 541–57.

Wilbanks, T. J. 2002. Scaling Issues in Integrated Assessments of Climate Change. In J. Rotmans and M. van Asselt, eds. *Scaling Issues in Integrated Assessment*: 19–49. Lisse: Swets and Zeitlinger.

Wilbanks, T. J., and R. W. Kates. 1999. Global Change in Local Places: How Scale Matters. *Climate Change*, **43**: 601–28.

World Resources Institute, United Nations Environment Program, United Nations Development Program, and the World Bank. 1998. *World Resources 1998–99.* New York: Oxford University Press.

2

The research strategy: linking the local to the global

Thomas J. Wilbanks, Robert W. Kates, David P. Angel, Susan L. Cutter,
William E. Easterling, and Michael W. Mayfield

In the beginning . . .

The Global Change and Local Places project of the Association of American Geographers originated in a 1992 meeting at which participants formulated three propositions:

- The grand query regarding the ways scale matters in understanding global climate change would benefit from detailed case studies of localities that were linked to scholars active in climate change-related research at global and national scales;
- Such case studies could constitute a basis for designing a research protocol for use in other local case studies, thereby helping build a body of empirical research that could serve as a basis for developing a bottom-up paradigm for global climate change research to complement the dominant top-down paradigm; and
- These locality studies should be based at universities whose faculty possessed detailed, long-term knowledge of their local areas, in some cases engaging scholars in global change research who might otherwise not normally participate in a large-scale research project.

Funding for the project outlined at the 1992 meeting was sought and eventually obtained from the National Aeronautics and Space Administration's Mission to Planet Earth Program (subsequently renamed Destination Earth). Intensive work on the project began in 1996 and continued through 2001. The several rounds of proposal writing that preceded funding refined the theoretical rationale for the project and its central components: four study areas located in Kansas, North Carolina, Ohio, and Pennsylvania; and three cross-cutting modules devoted respectively to estimating local greenhouse gas emissions, understanding the forces driving those emissions, and assessing local emission reduction potentials.

The Global Change and Local Places approach

Global Change and Local Places was envisioned as an experiment to determine whether examining climate change parameters at locality scale would improve the understanding of global climatic changes at broader scales, including global issues. The project was conceived only a short time before local and regional perspectives on global climate change became a central concern in the first national assessment of potential consequences of climate change in the United States, which triggered a new level of awareness of the importance of local-scale knowledge.

Conceptually, the project responds to research needs articulated in the early 1990s by the National Academy of Science – National Research Council Committee on Human Dimensions of Global Change (Stern *et al.* 1992; United States Global Change Research Program 1994). The project simultaneously addressed two basic questions: (a) the nature of global change, and (b) the scale at which it best could be observed and analyzed. In other words, in what ways are local events and processes something other than reduced-scale copies of those that occur at national, continental, and global scales? By focusing on such questions, the research team hoped to augment understanding of the links between global climate processes operating at macro-scales and significant processes that operate at micro-scales. The operating principles that guided the project are that:

• Scale matters;
• Local actions serve as foundations for global trends; and
• Local studies can be linked to larger global change models.

Though adopted as axioms for the project, these were also propositions to be examined, tested, and either supported, rejected, or refined during the course of the work.

Methodologically, the project team set out to design and test a common protocol for comparative research at locality scale that incorporated the dynamics of human activities and land uses, their driving forces, and the potential of local decision makers to alter greenhouse gas and aerosol production. This proposed protocol would find ways to cope with differences among places in availability of data (especially for reliable time series) and the variability in resources available to support local area research. A common protocol for studying localities could be formulated relatively easily by downscaling existing methods, but project members sought to understand environmental change more than climate change, and global change more than environmental change. Those priorities implied a clear need to integrate major ongoing changes in population growth and migration, economic restructuring, and technologies and their uses, with environmental processes. Such integration is difficult at best and may be more tractable locally than for larger areas.

Institutionally, Global Change and Local Places set out to link the detailed knowledge of localities possessed by faculty in universities having a strong commitment to their home areas with the knowledge of global-scale processes resident in the global change research community. In particular, the project sought to demonstrate that such universities – whether recognized as major research centers or not – possess large stores of local expertise and can contribute much that should be known to long-term monitoring and analysis of the human dimensions of global change at local and regional scales.

The four localities

At the outset, the Global Change and Local Places research team set out to test these propositions in three United States localities, each sized at a geographic scale of approximately 1° of equatorial latitude and longitude, or an area of about 110 kilometers (69 miles) on a side. That size was selected in part to nest within the then-prevalent 10° and the emerging 5° scales of disaggregated climate change modeling, and in part to assure some degree of land use diversity within each area. The three study areas were selected to offer a variety of conditions and processes related to environmental change from industry to agriculture and forestry, and from economic vigor to economic stagnation.

The initial three study areas were:

- a part of Southwestern Kansas with an economic base of integrated agriculture and live-stock production dependent upon irrigation from the Ogallala Aquifer;
- a portion of Northwestern North Carolina that included parts of both the Piedmont and the Blue Ridge, exemplifying economic change along the forested Appalachian front in the Southeast; and
- a segment of Northwestern Ohio, a part of the Great Lakes manufacturing belt recently reshaped by the decline of heavy industry in a part of the country long modified by urban–industrial settlement.

Early in the project, a fourth site was added to take advantage of ongoing research at the Pennsylvania State University: a part of the Susquehanna River basin centered on State College, Pennsylvania, where Penn State University is situated. This fourth area hosts a complex of a long-standing coal mining industry, mixed agriculture, and new technologies and practices spun off from a major state university. The original research design envisioned a total of nine study areas, but the funds obtained were sufficient for no more than three study areas. The four areas represent only a few of the many landscape and economy combinations extant in the country, but they embraced diversity adequate for project purposes.

Southwestern Kansas

The six counties of the Southwestern Kansas study area lie in the center of the High Plains of the United States and in the heart of the 1930s dust bowl (Figure 2.1). The region is flat, arid, and sparsely populated, supporting many more animals than people (Figure 2.2). Its major towns are Garden City, Dodge City, and Liberal. The region's complex, integrated pattern of agricultural production and processing is distinctive, especially its large beef exports to other areas and its reliance since the 1950s on irrigation from the Ogallala Aquifer. The area is distant from major population centers, and its residents tend to view themselves as independent and are skeptical of external interference. Only recently overwhelmingly Anglo in ethnic composition, the region's population now includes large numbers of recent arrivals from Latin America and Asia who have come for jobs in agricultural processing facilities.

Northwestern North Carolina

The twelve counties of the Northwestern North Carolina study area include a part of the Blue Ridge section of the Appalachian Mountains and a part of the rolling Piedmont lying between the Blue Ridge and the Coastal Plain (Figure 2.3). The mountain portion is characterized by a healthy tourist industry and forest product production, but the area's economy is dominated

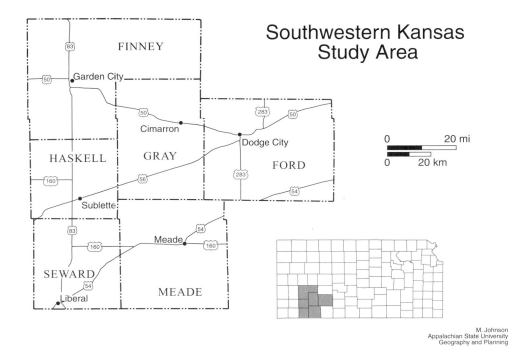

Figure 2.1 The Southwestern Kansas Global Change and Local Places study area.

Figure 2.2 Typical Southwestern Kansas study area residents.

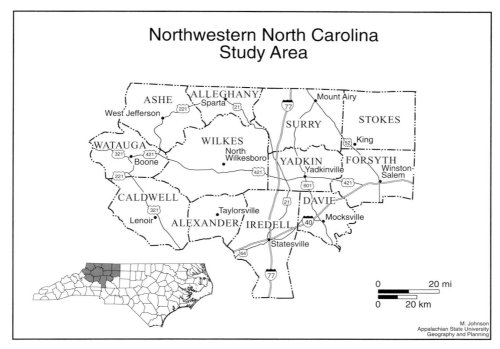

Figure 2.3 The Northwestern North Carolina Global Change and Local Places study area.

in value by small-scale manufacturing and agricultural production on the Piedmont, with Winston–Salem as the major center. In this portion of the region, the furniture and electronics industries have been expanding whereas textile production – one of the earlier industrial bases – has declined. The region's well-developed transportation network enables rural dwellers to work in urban industrial jobs. The region's characteristic border–South culture is being leavened by an influx of new residents from other areas, attracted by pleasant settings and a sense of economic potential (Figure 2.4).

Northwestern Ohio

Twenty-three counties of Northwestern Ohio constitute the third study area (Figure 2.5). Situated in the core of the nation's industrial heartland, Chicago lies to the west and Detroit a short distance north, and Cleveland and Pittsburgh sit to the east and southeast, respectively. The city of Toledo is the study area's major metropolitan center, a node for both rail and lake transportation. The study area also includes some areas of agricultural production and their urban service centers. The region's economy is dominated by manufacturing, but it and the larger Rust Belt of which it is a part have experienced widespread industrial restructuring in the past three decades (Figure 2.6). The new industries that have replaced the older iron, steel, and chemical bases pollute less than their predecessors. Culturally, the area exhibits the diversity characteristic of North American urban–industrial centers.

Figure 2.4 Northwestern North Carolina Study area golf course and resort.

Local politics focus on issues of economic viability and job creation, giving the area's major employers considerable political influence.

Central Pennsylvania

The five counties of the hilly to mountainous Central Pennsylvania study area have historically relied on coal and the main campus of The Pennsylvania State University for

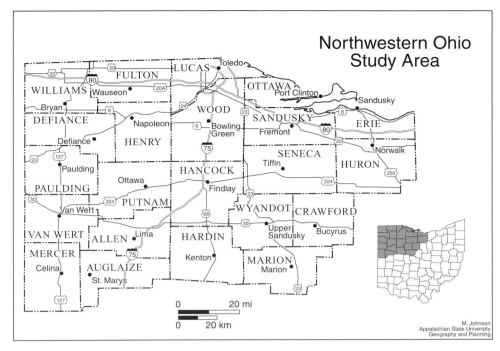

Figure 2.5 The Northwestern Ohio Global Change and Local Places study area.

Figure 2.6 Abandoned Autolite Spark plug plant in Northwestern Ohio study area.

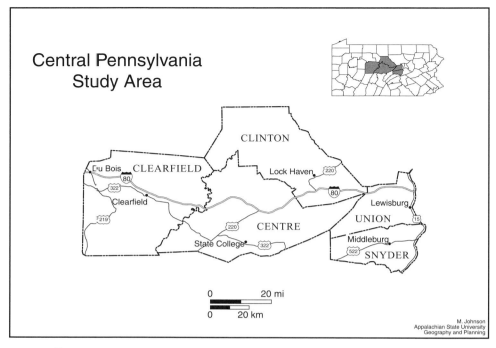

Figure 2.7 The Central Pennsylvania Global Change and Local Places study area.

livelihood (Figure 2.7). Penn State is located in State College, the area's largest settlement. This study area is a part of Appalachia, with the areas outside State College exhibiting many of the economic and social characteristics of that broader region: a historic dependence on mining, logging, and small-scale agriculture. Penn State's presence and continued growth has fostered high-tech manufacturing and numerous service industries in and near State College, but coal mining continues to be more important in the region than it is in the state or the nation. There exist sharp contrasts between the conservative, traditional cultures of most counties (some hosting Amish and Mennonite enclaves) on the one hand, and the diversity and liberal attitudes of the large university community on the other (Figure 2.8).

The three modules

To test the Global Change and Local Places propositions regarding scale, local action, and their connections to global models, Global Change and Local Places investigators asked several key questions in all four study areas:

- What have been the magnitudes and trajectories of local changes in greenhouse gas emissions and their sources and sinks since 1970, and what are they likely to be through 2020?

Figure 2.8 Amish farm in the Central Pennsylvania study area.

- What forces have driven these emission trajectories in the past, and what forces are likely to drive them in the future?
- What local capabilities exist to moderate changes in driving forces, are there local propensities to undertake such adaptive behavior, and how might possible policy mandates affect adaptation?

The means used to answer these questions were three research modules derived from the elements of integrative research on global change issues called for by the United States Global Change Research Program in 1994 (Figure 2.9). In the first module, changes in trace gas emissions, aerosols, and surface reflectivity were measured in 1970, 1980, and 1990 and linked to changes in their proximate sources. Projections for the same variables were made for 2000, 2010, and 2020. In the second module, changes in proximate sources would be linked to changes in driving forces through the use of explanatory models, again at ten year intervals from 1970 through 2020. The third module employed local knowledge, surveys, and analogs to assess capacities for mitigative action at the local level. Excluded from the Global Change and Local Places schema were local *impacts* of global climate change. At the time the project began, capabilities for producing reliable impact estimates for small areas did not exist. Were a similar project being

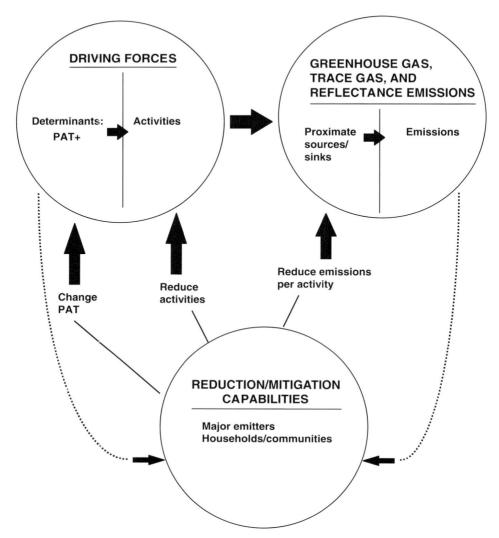

Figure 2.9 The Global Change and Local Places concept.

designed today, of course, impact estimates would most likely be included as a fourth module.

Estimating local greenhouse gas emissions

For each study area, emissions of greenhouse gases were estimated at county levels and aggregated to study area totals. Other aerosols and land cover reflectivity (albedo) were also considered as appropriate, on a case-by-case basis. The emissions and albedo were linked to local sources or sinks, such as power plants, transportation, industry, agriculture, and forestry. Estimated at ten-year intervals since 1970, the emission data were then compared with similar data for states and the nation.

Propositions associated with the initial design and execution of the estimation module were:

- Trace gases, aerosols, and reflectivity observed at locality scale will exhibit greater variation and volatility than has been measured globally;
- Differences will vary temporally as well as geographically; and
- These variations have critical policy implications for mitigating and managing greenhouse gases.

Understanding driving forces

Emission patterns were examined to understand the forces that drive the human activities which produce them, building in part on the **I = PAT** (**I**mpacts are a function of **P**opulation, **A**ffluence, and **T**echnology) identity (Chapter 1) as a heuristic. Other driving forces specific to each area were also considered by each local investigating team, with special attention given to the balance of endogenous and exogenous forces affecting each region. In almost any sizable region imaginable, some greenhouse gas emissions originating within the region are driven by demand for products and services located outside the area, and vice versa. External consumers use electricity generated by a major power plant located in the Northwestern North Carolina area, for example, while local natural gas consumption results in methane emissions at wellheads outside the area. Because analysis of driving forces directs attention to demand rather than to the point sources that meet it, the driving forces module incorporated estimates of a second set of demand-based estimates of locally-induced emissions as a basis for comparison.

Initial propositions regarding driving forces were:

- The I = PAT formulation helps identify proximate (but not ultimate) driving forces;
- Locality-scale analysis makes it possible to separate the myriad processes underlying the I = PAT identity;
- Greenhouse gases can be separated with respect to geographically endogenous and exogenous driving forces; and
- At locality scale, I = PAT must be expanded (I = PAT+) to account for the effects of linkages to areas beyond each locality's boundaries.

Exploring emission reduction and mitigation potentials

Beyond both emissions and their driving forces lie capacities for reducing greenhouse gas emissions, locally and more generally. The third Global Change and Local Places module considered potentials for reducing emissions through local actions, local propensities to undertake such actions, and the constraints that might hamper propensities and reductions. Surveys of the attitudes of major emitters and the general populations within the study areas were buttressed by examinations of local behavior in past instances of analogous environmental challenges. Global Change and Local Places analysts also looked beyond recent and current potentials toward possible future (to 2020) adaptation and mitigation based on emission projections, conceivable emission reduction targets, realistic emission reduction alternatives, and plausible conditions under which such alternatives might be adopted. A central issue was understanding which factors might be amenable to local action as contrasted with actions that would have to be taken at corporate, state, and national scales.

Propositions incorporated at the outset of the project were:

- Most mitigation will be the result of local action;
- Previous environmental actions can serve as analogs and indicators of propensity to mitigate, as can direct surveys and focus group results; and
- National- and state-mandated mitigation efforts have significantly different effects in different localities.

Research methods for learning about localities

To assure comparability across the four sites, the Global Change and Local Places team developed a set of methods designed for use at the local scale (about 12,300 square kilometers or 4,750 square miles). This tool kit included methods to estimate greenhouse gas emissions from the four local areas, methods to identify the forces driving those emissions, methods to assess local attitudes toward emission abatement, and the general approach for evaluating local capacities to mitigate greenhouse gas emissions. Close attention to methods development proved necessary because the Global Change and Local Places team quickly learned that the widely used methods developed for estimating greenhouse gases at national and global scales could not be applied readily at locality scales (Chapter 12).

Estimating greenhouse gas emissions from local areas[1]

The major greenhouse emission producers in all four study areas are related to fossil fuel production and consumption (manufacturing, electricity generation, transportation, and household heating); to forestry and agriculture (livestock, wetlands, fertilization, land clearing, and timber production); to waste disposal (landfills and incineration); and to the manufacture and use of ozone-depleting chemicals. Trace greenhouse gases resulting from these activities are carbon dioxide, methane, nitrous oxide, and ozone-depleting chemicals that accrue in the atmosphere where they increase positive radiative forcing and thereby warm the Earth's surface.[2] The same industries and consumption also emit sulfate aerosols and other airborne particles that decrease positive forcing and thus counter greenhouse warming. Local changes in land cover lead to changes in reflectivity (albedo) that can also affect radiative forcing. Taken together, factories, vehicles, households, fields, forests, streets and

---

[1] This section is adapted in part from Easterling *et al.* (1998).

[2] Most inventories track both ozone-depleting chemicals and their substitutes. Ozone-depleting chemicals include chlorofluorocarbons and hydrochlorofluorocarbons, and ozone-depleting substitutes include hydrofluorocarbons, perfluorocarbons, and sulfur hexafluoride.

Table 2.1 Principal sources and sinks of Global
Change and Local Places greenhouse gases

Gas	Sources and Sinks
CO_2	Fossil fuel combustion
	Natural gas flaring
	Cement manufacture
	Lime manufacture
	Limestone and dolomite use
	Soda ash manufacture and consumption
	CO_2 manufacture
	Land-use change and forestry (**Sink**)
CH_2	Stationary sources
	Mobile sources
	Coal mining
	Natural gas systems
	Petroleum systems
	Petrochemical production
	Silicon carbon production
	Enteric fermentation
	Manure management
	Rice cultivation
	Agricultural residue burning
	Landfills
	Wastewater treatment
N_2O	Stationary sources
	Mobile sources
	Adipic acid
	Nitric acid
	Manure management
	Agricultural soil management
	Agricultural residue burning
	Human sewage
	Waste combustion

Source: United States Environmental Protection
Agency (1998).

roads, and animals constitute billions of point sources of emissions, aerosols, and land cover
changes that must somehow be estimated.

Greenhouse gas emission inventory methodology

Most if not all greenhouse gas emission inventories account for emissions on both by gas
and by sector bases (Table 2.1), corresponding to the reporting standard established by the

Intergovernmental Panel on Climate Change and adopted by the United States Environmental Protection Agency for its national and state-level emissions inventory protocols (United States Environmental Protection Agency 1995a,b). The major gases and sectors tracked in inventories are aggregated into a common measure equivalent to the major gas (carbon dioxide) or its carbon content using one of two general equations:

(activity data) × (emissions factor) × (Global Warming Potential)
    = emissions of carbon dioxide equivalent,

or

(activity data) × (emissions factor) × (Global Warming Potential) × (12/44)
    = emissions of carbon equivalent.

Activity data refers to the measure of a process emitting a greenhouse gas as a byproduct. Examples include gallons of gasoline combusted, tons of lime manufactured, and pounds of fertilizer applied. The *emissions factor* is the empirically derived ratio of greenhouse gas emitted per unit of activity. In some cases this factor is known precisely; in other cases only a range of possible values is known. Global Warming Potential converts units of other gases to units of carbon dioxide in terms of equivalent radiative forcing. The (12/44) term converts units of carbon dioxide to units of carbon by counting the molecular mass of carbon as a proportion of carbon dioxide.

Carbon dioxide equivalents are useful because carbon dioxide is the most prevalent of the greenhouse gases, and carbon equivalents are helpful for tracking stocks and flows of greenhouse gases among sources and sinks in the global carbon cycle. Various inventories have used both measures. In the Global Change and Local Places project, local inventories are usually reported in carbon dioxide equivalents. Carbon equivalents are sometimes used in making scale comparisons, given original sources.

Global Warming Potential (GWP) requires further elaboration. Radiative forcing varies considerably among trace gases and with the time the gas resides in the atmosphere, that is, the time from emission into, to removal from the atmosphere). For example, a molecule of methane is more than twenty times more efficient than a molecule of carbon dioxide in absorbing infrared radiation, calculated over a one hundred year period of residence in the atmosphere. It is helpful from both scientific and policy perspectives to develop an equivalence scale for various trace greenhouse gases, and for such purposes carbon dioxide was chosen as the reference gas and 100 years as the atmospheric residence time (Intergovernmental Panel on Climate Change 1995). Global warming potential incorporates these benchmarks, and thereby provides a basis for comparing the combined direct and indirect radiative forcing of different levels of emissions of different gases (Box 2.1).[3]

---

[3] Greenhouse warming potential values are subject to revision to reflect improved estimates of the relevant atmospheric chemistry. The global warming potential values used by Global Change and Local Places are not necessarily the most recent ones available. They are as follows (United States Environmental Protection Agency 2001): carbon dioxide = 1; methane = 21; and nitrous oxide = 310.

Box 2.1 Global warming potential

The ratio of both direct and indirect radiative forcing of a gas compared to carbon dioxide is called its global warming potential (GWP). By definition, the global warming effect of carbon dioxide is set equal to 1, and different values of global warming potential for different greenhouse gases reflect their respective cumulative net radiative forcing over a specified time period, which is 100 years for Intergovernmental Panel on Climate Change and United States protocols, and also the period used in Global Change and Local Places analyses. Global warming potential values are revised from time to time in response to improvements in relevant atmospheric chemistry. The global warming potential values used by Global Change and Local Places (not necessarily the most recent now available) are: carbon dioxide = 1; methane = 21; and nitrous oxide = 310 (Appalachian State University 1996).

A first attempt at greenhouse gas emissions accounting at the sub-national level resulted from United States Environmental Protection Agency financial and technical assistance to states to compile comprehensive greenhouse gas inventories, state action plans, and innovative demonstration projects. Environmental Protection Agency methods are based on the approach established by the Intergovernmental Panel on Climate Change (1992, 1995). The Intergovernmental Panel on Climate Change methods were designed to be used in countries throughout the world and with a wide range of input data. The Environmental Protection Agency has produced a 325 page *State Workbook* (United States Environmental Protection Agency 1995b) with instructions for calculating emissions of carbon dioxide, methane, nitrous oxide, and some of the ozone-depleting compounds (HFC-23 and PFCs). Because detailed data concerning energy-consuming activities are available for the United States, the Environmental Protection Agency methods are more comprehensive than those established by the Intergovernmental Panel on Climate Change (United States Environmental Protection Agency 1995b, 1996).

Unlike many air pollutants, greenhouse gas emissions are not measured directly, but are inferred either from the use of materials that yield such gases or from processes that produce them. The *materials* approach is exemplified by the calculation of carbon dioxide released from the combustion of coal in power plants. Data on coal consumption by utilities exists at the state level, so simply multiplying those values by coefficients that account for the carbon content of the coal and the combustion efficiency of the boilers suffices. The *process* approach is exemplified by the calculation of carbon dioxide emissions from a cement plant. The amount of carbon dioxide released during the calcining process can be estimated if the emissions of carbon dioxide per ton of cement produced and the cement output of the plant are known (United States Environmental Protection Agency 1995b).

Such coefficient-based methods result in state inventories of greenhouse gas emissions that are summarized in matrix form showing different gases in tons per year originating from sources that include both materials (fossil fuel and biomass) and activities (specific industrial production processes, agriculture, and waste disposal). Emissions are further subdivided by source (commercial, industrial, residential, utilities, transportation, and fuel production and distribution). All the emissions are then converted to standard units of equivalent tons of carbon dioxide as described above.

Adapting greenhouse gas inventory methods to Global Change and Local Places

The Global Change and Local Places team estimated emissions of carbon dioxide, methane, and nitrous oxide. Two classes of gas were excluded from the Global Change and Local Places inventories. Data limitations regarding the number and nature of activities at locality scale that emit ozone-depleting compounds proved insurmountable, forcing the team to omit them. Because one hundred year greenhouse warming potential values for ozone-depleting compounds range between 100 and 25,000 with most registering well over 1,000, including them would significantly alter total carbon dioxide equivalent emission estimates for regions that emit large quantities of these gases. At present, however, the magnitude of carbon dioxide equivalent emissions due to ozone-depleting compounds is relatively small, and emission policies stemming from the Montreal Protocol are aimed at further reductions.[4] The second class of gases not included in Global Change and Local Places inventories is *criteria pollutants* (carbon monoxide, oxides of nitrogen, and volatile organic compounds). Because these gases contribute only indirectly to the greenhouse effect, they do not have global warming potential values and are not normally incorporated in inventories such as those constructed by Global Change and Local Places researchers. The United States Environmental Protection Agency handbook (1995a) contains details regarding ozone-depleting compounds and criteria pollutants.

Emissions data were tabulated by major source category for 1990 in all four Global Change and Local Places study areas and those data were the basis for estimates of 1970 and 1980 emissions. For example, transportation emissions in 1990 were adjusted to yield 1980 and 1970 estimates by considering changes in three factors: the number of registered vehicles, average fuel efficiency, and the average vehicle miles traveled. Estimates of 1980 emissions were calculated using the formula:

(1980 emissions) = (1990 emissions) × (registered vehicles in1980/registered vehicles

1990) × (miles traveled per vehicle in 1980/miles traveled per

vehicle 1990) × (miles per gallon 1990/miles per gallon 1980).

Practicality dictated an assumption of constant emission factors in activity data between time periods, though assuming, for example, that a gallon of gasoline burned in 1980 emits the same amount of carbon dioxide as in 1990 almost certainly introduces some error into the estimates.

---

[4] For example, including ozone-depleting chemicals here would have increased overall 1990 carbon dioxide equivalent emissions by about 5% in the Central Pennsylvania study site, and about 2% in the United States as a whole.

Data sources for Global Change and Local Places greenhouse gas emission estimates[5]

Source categories utilized in Global Change and Local Places reflect the major categories used in United States emissions inventories. They include:

- Fossil fuel combustion;
- Biomass combustion;
- Production processes;
- Agriculture;
- Waste disposal; and
- Land-use change and forestry.

An example of a greenhouse gas emission inventory for a Global Change and Local Places local study area is shown in Table 4.1 (pp. 84–5).

Not surprisingly, data for small areas are more difficult to obtain than at higher levels of aggregation. Data are not collected, are withheld as proprietary, or are made available only at great expense. When available, they often do not match the area being studied and must be allocated to it with much difficulty. Examples abound: data on fossil fuel consumption, the major single variable in emissions, are not available locally; electricity generation and consumption data rarely fit local areas, and utilities are becoming less willing to share such information when they do have it; and estimates of ozone-depleting compounds are based on national production data.

Consider the fossil fuel consumption data. In both the United States and Canada, detailed energy commodity statistics are available at the state and provincial levels, but such data are not generally collected or compiled for places such as municipalities or counties, for which other sources and methods must be used. A common method is to simply apportion national, state, or provincial level data to cities or counties in proportion to their respective populations or economic activity, but this method is inconsistent with the central focus of Global Change and Local Places: measuring and understanding local variability. Like other groups such as the International Council for Local Environmental Initiatives Cities for Climate Protection Program (Chapter 11), Global Change and Local Places utilizes a combination of bottom-up estimates and top-down normalization to characterize local emissions in ways that yield insights into the dynamics of greenhouse gas emissions at local scales.

A particular challenge is quantifying transportation-related emissions at a local scale. Transportation-related emissions account for nearly one third of all greenhouse gas emissions in the United States (United States Environmental Protection Agency 1995a), but can

---

[5] Supporting data and results for Global Change and Local Places research were generated by the respective study sites. For the Northwestern North Carolina site, see Appalachian State University (1996), Mayfield, *et al.* (1996), Lineback *et al.* (1999), and DeHart and Soulé (2000). For the Southwestern Kansas site see Harrington *et al.* (1997), Holden *et al.* (1997), and Harrington (1998). For the Northwestern Ohio site see Muraco and Attoh (1997a, b), Attoh and Muraco (1997), Lindquist (1997), and Reid (1997). For the Central Pennsylvania site see Denny (1999), and United States Environmental Protection Agency (1997).

comprise more than half the greenhouse gas emissions in some urban areas. Data making it possible to calculate carbon dioxide emissions directly from gasoline or diesel fuel consumption or sales do not exist at the county level in any of the Global Change and Local Places study sites and appear to be available only for some states in the United States.

In states where accurate fuel sales data are available at the county level, significant boundary issues must be considered. Fuel sales in a jurisdiction are not necessarily accurate indicators of fuel consumption in the same area: transportation fuel purchased in one jurisdiction may be burned elsewhere and transportation fuel burned in the jurisdiction may have been purchased elsewhere. Retail gasoline and diesel sales data are often available from private sources, and they are available down to postal code and address levels, but the issue remains of the mismatch between fuel sales and fuel consumption data. Further, retail sales of transportation fuels do not include the considerable portion of transportation fuel sold at wholesale for use by government and corporate vehicle fleets.

In the Global Change and Local Places analyses, the major surrogates for carbon dioxide generated from local fuel sources are local emission estimates of carbon monoxide air pollution data, which in the United States are available for local areas. More specifically, Global Change and Local Places uses monitored or modeled carbon monoxide (a legally defined pollutant) as a surrogate for carbon dioxide, based on the assumption that for specific sources or sectors, the emission of carbon dioxide is proportional to the emissions of carbon monoxide. Carbon monoxide emissions are estimated or calculated for individual point sources – industrial plants, commercial–institutional users, and power plants – and these emission data are reported directly to the Environmental Protection Agency (United States Environmental Protection Agency 1996). These specific point sources were used to estimate industrial, commercial and utility emissions. In the cases of the residential- and transportation-based carbon monoxide emissions, which arise from hundreds of millions of sources, estimates are made on the basis of models of emission rates from different categories of household or vehicle. For transportation, Global Change and Local Places used readily available data – vehicle miles traveled by county – as input into a model that estimated carbon monoxide and other vehicle emissions (United States Environmental Protection Agency, Office of Transportation and Air Quality 2002). The county estimates of carbon monoxide were then used to apportion state fuel-sales-derived carbon dioxide estimates to the county level.

There are other difficulties at all scales. For example, emissions of methane during the production and distribution of natural gas come from a variety of sources, including field production, processing plants, storage facilities, transmission facilities, compressor stations, and distribution networks. Greenhouse gases are released at every stage in the production and delivery of natural gas through combustion of fuels, flaring, fugitive emissions from leaks, and pipe evacuation for maintenance. There is greater uncertainty about the amounts of methane released in natural gas production and delivery than in any other sector of emissions. Work in the Southwestern Kansas Global Change and Local Places study area indicates that emissions from pipeline purging and fugitive emissions are significantly higher than the United States Environmental Protection Agency protocols indicate.

Another problem for the Global Change and Local Places team is how to conceptualize the use of biomass as fuel. Intergovernmental Panel on Climate Change/Environmental Protection Agency protocols include an emissions inventory category for land-use change. If biomass is harvested and combusted faster than it is replaced, then the net carbon flux from the imbalance is recorded in the land-use change category. Emissions are reported only if there is a net positive difference between carbon dioxide emissions from biomass combustion and carbon dioxide taken up by changes in land use or land cover. Hence, carbon dioxide emissions from biomass combustion are not included as such in the combustion-related emissions, but are embedded in carbon emissions reported from land-use change. Biomass combustion can and does make a large contribution to energy supply and thereby displaces what would otherwise be fossil fuel combustion, however. Therefore, local methods for greenhouse gas emissions analysis should track carbon dioxide emissions from biomass combustion (Chapter 4), but omit biomass from the inventory total to avoid double counting.

Identifying the forces driving greenhouse gas emissions from local areas[6]

The Global Change and Local Places analysis of driving forces began with the set of greenhouse gas emissions defined above, represented in terms of total global warming potential, involving direct emissions from all proximate sources within each of the study areas. These data do not, however, include embodied or indirect emissions that may be associated with these proximate sources but that are not released in the study area, such as emissions generated outside the study area in the manufacture of an automobile owned and used inside it, or emissions inside the study area from vehicles passing through it on an interstate highway. An alternative to this strictly geographical accounting system would assign all emissions to the activities with which they are directly or indirectly associated. For instance, Morioka and Yoshida (1995, 1999) sought to determine the greenhouse gas emissions for which an individual household is responsible, both directly (household heating for example) and indirectly (power plant emissions caused by household electricity use, for example). Global Change and Local Places created two sets of emission inventories, one using the categories of the United States Environmental Protection Agency state inventory and one focused on groups of end users of emission-generating products.

Estimating local emissions offers conceptual difficulties. Some emissions, such as burning fossil fuels for heat or for motive power, occur at the points of final consumption. Other emissions are separated in space or time from the points of final consumption that cause the emissions. Electric power plant emissions and oil and gas production and transmission emissions may be far removed from the sites of final consumption of the power or fuels that engender them. Organic waste disposal by businesses and households today results in future methane emissions, perhaps from landfills hundreds of kilometers away. Ultimately, all greenhouse gas emissions could be linked to final consumption, but the extent to which

[6] This section is based on Angel *et al.* (1998).

such a principle should and can be applied is a central question in designing local emission inventories. Those who would inventory a locality's emissions must decide when and how to allocate emissions that are produced outside the area for final consumption within it, and how to distribute emissions within the area.

In principle, final consumers should always be held accountable for the full cycle of emissions their activities generate. In practice, many difficulties of data conceptualization, collection, and double counting arise in attempting to attribute all emissions to final demand. The practical principle of allocating emissions in accordance with the ability to reduce them can be helpful in determining the most useful method. The two most important kinds of off-site carbon dioxide emissions are those arising in the production and transportation of fossil fuels and those caused by the production and transportation of electricity consumed locally but generated at a distant site.

A *full fuel cycle emissions analysis* would allocate to the locations of final consumption all the emissions generated upstream from consumption – from oil well to gas tank in the case of petroleum or from gas field to home furnace in the case of natural gas – in addition to the emissions at the point of use. Typically, the resulting full cycle emission coefficients are about 10–15% higher than point-of-end-use emissions estimates. Global Change and Local Places analysts did not attempt full cycle emissions coefficients in their inventories, in order to avoid double counting.

End user analysis

Within study areas, Global Change and Local Places staff sought to allocate emissions in ways that would make it possible to inform local economic enterprises, governments, institutions, and householders of their roles in creating emissions and of their capacities to reduce them. More specifically, for the purposes of the Global Change and Local Places project it was desirable to be able to ascribe emissions to those actors who have the capacity and authority to change those emissions. Unfortunately, none of the standard accounting frameworks tracks authority over emissions or the distribution of benefits associated with them. Standard international, national, and even state inventories cannot be disaggregated readily for that purpose because they mix materials and activities, as well as producers of emissions and consumers of emission-generating products. Household emissions, for example, are listed under as many as five different categories: electricity consumption is estimated within a category of utilities; household travel emissions within a transportation category; household fuel consumption for heating within a residential category; household waste within waste in municipal landfills; and household ozone-depleting compounds, if estimated at all, are prorated from national production figures.

To move toward an inventory reflecting agency as well as location, Global Change and Local Places analysts reallocated county level greenhouse gas emissions (with the exception of ozone-depleting compounds) to the *end user categories* of agriculture, residential, and industrial–commercial. The latter collapsed commercial and industrial into a single category because of the absence of separate data. To illustrate, in order to inform households of their actual shares of local emissions, all of the household-related portions of the following emission sources were included in the *residential* end user category: waste incineration, sewage, landfills, fossil fuel use, biomass, electrical utility, non-highway vehicles, and highway vehicles.

This reallocation process required major changes in tabulating emissions by electric utilities. In three of the four study areas, electric power plants were the largest single point source of greenhouse gas emissions. Opening or closing a single coal-fired power plant in a relatively small locality has a dramatic impact on total greenhouse gas emissions from that particular geographic unit. Conversely, the availability of nuclear electricity (as in Northwestern Ohio) can greatly affect estimated emissions from end user groups. For these reasons, analyses of driving forces included in end use inventories only the emissions associated with electricity consumed within study areas, excluding the emissions from electricity generated locally but used outside the study areas. Estimates of emissions from electricity generated outside the study area but consumed inside the area *were* included in the inventory, in effect allocating electricity generation emissions to the places where the electricity is used rather than where it is generated.

Empirical decomposition

With emissions data by end use category in hand, the challenge Global Change and Local Places researchers faced was to specify the chain of decisions, events, and processes that give rise to changes in greenhouse gas emissions. For example, an increase in net emissions from households in the Northwestern North Carolina study area between 1970 and 1990 (Angel *et al.* 1998) (Chapter 8) results from increases in the number of households, the number of vehicles they registered, and the amount of electricity they consumed, but the increases were offset by improved vehicle fuel efficiency, a decrease in household fossil fuel use, and fuel switching in electricity generation. Such empirical decomposition of change in household energy emissions or emissions for other user groups is largely determined by data availability and conceptual clarity. Where data are available, driving forces can be empirically decomposed to reveal an entire explanatory chain and identify the most important variables, but this analytical process is limited to the variables specified.

The six variables linked to household emissions can also be broadly linked to the I = PAT identity variables of population, affluence, and technological change as driving forces. Household growth is driven by population, increased vehicle and electricity use, affluence, fuel switching, increased fuel efficiency, and decreased fossil fuel use resulting from technological changes, but only partly so. For example, growth in the number of households is a function of changing patterns of marriage and family as well as of population growth. Marriage and family patterns are complex phenomena driven by causes not embraced by the I = PAT identity. Unraveling the causal chains underlying changes in greenhouse gas emissions is a formidable task, conceptually as well as empirically.

Even so, conducting analyses of driving forces at locality scale provides insights beyond the I = PAT identity by enabling investigators to examine driving forces in the social science tradition of using case study research to identify causal pathways. Local scale makes it possible to examine local processes that shape outcomes and to deal more comprehensively with complex systems, and locality scale may yield special explanatory power as a result of the intersection of local, national, and global processes in the context of specific places.

Assessing attitudes toward greenhouse gas emission abatement in local areas

In order to gauge local attitudes toward greenhouse gas emission abatement, Global Change and Local Places investigators undertook two courses of action in the three original study areas. Although similar efforts were also pursued in the Central Pennsylvania study area, its different conceptual origins and sources of financial support required attitude surveys that are not directly comparable to those employed by the Global Change and Local Places project. After initial experiments with focus groups, the Global Change and Local Places project conducted a survey of local householders to reveal the range of general community attitudes toward global climate change. Then representatives of the major greenhouse gas emitters in each study area were interviewed informally to determine the attitudes held by those with the greatest potential to bring about changes in emissions.

The household survey

The Global Change and Local Places household questionnaire was designed to elicit residents' knowledge and opinions regarding energy use, greenhouse gases, and global climate change. Each study site began with a basic survey, which was then modified to suit its locality. The questionnaires contained four main topical sections: global change and greenhouse gases; local and household mitigation strategies; household energy use; and socio-demographic characteristics. Five questions were selected from national surveys of climate change attitudes and included to permit comparability with national opinions.

Questionnaires were mailed to random samples of households within each study area during the late summer and early fall of 1997. Global Change and Local Places's initial intent was to sample approximately 1% of the households in each study area county. That goal was realized in Southwestern Kansas, but it could not be met in Northwestern North Carolina and Northwestern Ohio because of the major urban centers located within those study areas. To maintain consistency in the number of surveys sent, 700–800 surveys were mailed in the Northwestern North Carolina and Northwestern Ohio study areas, representing about 0.25% of the total households in the combined regions.

The initial addresses were drawn from DeLorme *Phone Search USA* (1998) and the 1990 LINC data (1998) for total occupied households in Northwestern North Carolina. Using a modified Dillman (1978) method, the questionnaires were distributed. When responses were not received after four weeks, a reminder card was sent. A second questionnaire was mailed

Table 2.2 Responses and response rates for the Global Change and Local Places
household surveys

Potential Responses = Surveys Sent minus Undeliverable
Total Usable Responses = Surveys Returned minus (Refusals + Unusable)
Total Response Rate = Surveys Returned divided by Potential Responses
Total Usable Response Rate = Total Usable Responses divided by Potential Responses

	Northwestern North Carolina	Southwestern Kansas	Northwestern Ohio	Total
Number of Households	291,179	28,443	609,429	929,051
Surveys Sent	846	702	810	2,355
Undeliverable	159	118	138	415
Potential Responses	684	584	672	1,940
Surveys Returned	204	219	174	597
Refusals	6	16	0	22
Unusable	4	4	3	11
Total Usable Responses	194	199	171	564
Total Response Rate	29.8%	37.5%	25.9%	30.8%
Total Usable Response Rate	28.4%	34.1%	25.4%	29.1%

Table 2.3 Sectors represented in study area interviews with major emitters

Southwestern Kansas	Northwestern North Carolina	Northwestern Ohio
Electric power generation	Electric power generation	Electric power generation
Agriculture	Manufacturing	Steel manufacturing
Natural gas industry	Natural gas industry	Oil companies
Transportation	University	Automobile manufacturing

three weeks later if no response was received, with a follow-up postcard shortly thereafter, but no subsequent follow-up. Response rates ranged from 37.5% in Southwestern Kansas to 25.9% in Northwestern Ohio, with an average for all three sites of 30.8% (Table 2.2). The proportion of usable questionnaires was slightly lower, at 29.1% for all sites.

Interviews with major emitters

Households represent one group of major emitters of greenhouse gases through end use consumption. A second consists of electric power plants or industrial facilities that were identified in local emissions inventories as large proximate emitters. Face-to-face or telephone interviews were conducted with a limited number of individuals employed by the largest local contributors to study area greenhouse gas warming potential and with selected local leaders (Table 2.3).

The interviews were unstructured and conducted informally, and the specific questions asked depended on local circumstances. Four broad topics were used to focus the open-ended interviews:

- Level of concern and awareness regarding climate change. 'Do you believe global warming to be a real concern?'
- Whether the respondent had any specific strategies for decreasing greenhouse gas emissions. 'Are there any actions that you would consider taking in order to reduce greenhouse gas emissions?'
- Perceived (preferred) locus of control for greenhouse gas mitigation activities. 'Who should be most responsible for reducing greenhouse gases – individual households, businesses, local governments, state governments, or the federal government?'
- Technology awareness as a means of reducing emissions. 'Do you know of specific technologies that could be used to reduce greenhouse gas emissions?'

Evaluating capacities to mitigate greenhouse gas emissions from local areas

Global Change and Local Places analysts explored potentials for greenhouse gas emission reduction through local agency by posing two arbitrary emission reduction targets, assessing alternate ways to meet those targets, and estimating the likelihood of different causal factors related to such pathways. One question of particular interest was whether local knowledge and local opportunities for action might make emission reductions more likely, more attractive, or less costly, compared with generic broader-scale emission reduction policies and actions.

As described in more detail in Chapter 11, the general approach to modeling future emission reduction potentials in the four study areas was to:

- Construct three scenarios of greenhouse gas emission trends to 2010 and 2020;
- Determine how the three scenarios would have to change to meet two emissions reductions targets; and
- Consider alternative possibilities and probabilities for meeting those targets.

The three emissions scenarios were:

- Simple extrapolation of emission trends observed from 1970 to 1990;
- Estimates based on projections of demographic and economic conditions at county levels for 2010 and 2020 prepared by a commercial forecasting firm (Woods & Poole Economics 1998);
- An approximation of best local knowledge in forecasting greenhouse gas emissions for 2010 and 2020 based on modifications of the Woods & Poole estimates to fit local expectations of whether the best estimates for particular sectors are the low, medium, or high Woods & Poole forecasts for those sectors.

These unmodified and modified Woods & Poole projections of economic and demographic drivers were then converted into emissions estimates (Box 2.2).

Box 2.2 Using data from Woods & Poole Economics, Inc. to project future greenhouse gas emissions at a local scale

Projecting futures is one of the more difficult challenges in analyzing greenhouse gas emission reduction prospects for local areas. Linking emission futures to reliable economic projections at a local scale, however, provides a way for local groups to formulate such projections. In the Global Change and Local Places analysis, economic projections from Woods & Poole Economics, Inc. (1998) were the basis for projecting emissions for each of three user groups:

• Residential;
• Industrial and commercial; and
• Agricultural.

State profiles for 1998 for Kansas, North Carolina, Ohio, and Pennsylvania – the most recent available at the time of the analysis – contained detailed year-by-year county forecasts of economic data to 2020. Other projections were available, but Woods & Poole data were used because its county-level projection algorithms yield results that are consistent from state to state.

Three variables representing population and economic growth were used to project residential emissions at five year intervals from 1995 to 2020: total housing, total population, and per capita income in constant dollars. The mean projected change in these three variables was multiplied by baseline 1990 greenhouse gas emissions for each county. County emissions were summed to yield each study area's total projected residential emissions for each five-year period. Each study area's totals were then graphed as the medium scenario, with two parallel lines showing values 20% more than (the high scenario) and 20% less than (the low scenario) the medium values.

Similar methods were used to project emissions for the other two user groups. Commercial and industrial emission projections were based on value added from manufacturing, total employment (excluding farm employment), and total earnings (except farm earnings). Data on value added by manufacturing from the United States Census of Manufactures for 1972, 1982, and 1992 were extrapolated for use with Woods & Poole data for the other user group categories. Agricultural emissions were based on Woods & Poole projections of farm employment, farm earnings, and agricultural service earnings, again in constant dollars. The approach was developed and standardized by the Appalachian State University members of the Global Change and Local Places team in order to permit comparisons of greenhouse gas projections among the four study areas.

Two emission reduction targets were selected: 7% below the 1990 level, the United States emission reduction target defined in the Kyoto Protocol of the Framework Convention on Climate Change, and 20% below the 1990 level, a target advocated by many groups and adopted by some cities, and an example of what could emerge in the coming decade if concerns about reducing rates of greenhouse gas emission increases and resulting global climate change become more urgent.

Local area greenhouse gas emission sources and sinks were examined to identify sets of actions that would reduce local emissions by 7% and by 20% from their 1990 levels in two ways. Local area assessment teams used information from the local study and other local knowledge to identify what appeared to them to be the best local alternatives for reducing greenhouse gases. These local portfolios of actions were then compared with two generic alternatives:

- those used in the International Council for Local Environmental Initiatives's analysis for member cities of the Cities for Climate Protection campaign, and
- those that have been identified for greenhouse gas emission reductions at the national level, in such studies as the United States Department of Energy's *Scenarios of U.S. Carbon Reductions* (Interlaboratory Working Group 1997)

to see whether assessing mitigation potentials at locality scale produces different understandings than assessments at national scale. Finally, the portfolios of potential actions to meet emission reduction targets were evaluated in terms of the existence of feasible paths for actual implementation.

REFERENCES

Angel, D. A., S. Attoh, D. Kromm, J. DeHart, R. Slocum, and S. White. 1998. The Drivers of Greenhouse Gas Emissions: What Do We Learn from Local Case Studies? *Local Environment*, **3** (3): 263–78.

Appalachian State University. 1996. *The North Carolina Greenhouse Gas Emissions Inventory for 1990*. Boone, NC: Appalachian State University Department of Geography and Planning.

Attoh, S. and W. Muraco. 1997. Testing the Applicability of the IPAT Model in Northwest Ohio. Paper presented at the Meetings of the East Lakes Division of the Association of American Geographers, Michigan State University, East Lansing, MI, October, 1997.

DeHart, J. L. and P. T. Soulé. 2000. Does I = PAT Work in Local Places? *The Professional Geographer*, **52**: 1–10.

DeLorme *Phone Search USA*. 1998.

Denny, A. S. 1999. Greenhouse Gas Emissions from Coal Combustion in Central Pennsylvania: Addressing Vulnerability through Technological Options. M.S thesis, Department of Geography, The Pennsylvania State University.

Dillman, D. A. 1978. *Mail and Telephone Surveys: The Total Design Method*. New York: John Wiley & Sons.

Easterling, W. E., C. Polsky, D. Goodin, M. Mayfield, W. A. Muraco, and B. Yarnal. 1998. Changing Places, Changing Emissions: the Cross-Scale Reliability of Greenhouse Gas Emission Inventories in the US. *Local Environment*, **3**: 247–62.

Harrington, J. Jr., 1998. A Climatology of Changing Spring Dew Point Temperatures: Dodge City, Kansas. *Preprints Ninth Symposium on Global Change Studies*: 114–16. Boston: American Meteorological Society.

Harrington, J. Jr., D. Goodin, and K. Hilbert. 1997. Global Change in Local Places: Satellite Reflectance Data for Southwest Kansas. *Papers and Proceedings of Applied Geography Conferences*, **20**: 43–7.

Holden, G., J. Harrington, M. Goodin, and J. DeHart. 1997. Estimating Changes in Greenhouse Gas Emissions for the Natural Gas Industry in Southwest Kansas. *Papers and Proceedings of the Applied Geography Conferences* **20**: 20–24.

Intergovernmental Panel on Climate Change, Organization for Economic Cooperation and Development. 1992. *National Inventories of Net Greenhouse Gas Emissions: IPCC Guidelines for Preparation and Reporting*. Paris: OECD.

Intergovernmental Panel on Climate Change, Organization for Economic Cooperation and Development. 1995. *Greenhouse Gas Inventory Workbook*. London: United Nations Environment Program, The Organization for Economic Co-operation and Development, and the Intergovernmental Panel on Climate Change.

Interlaboratory Working Group. 1997. *Scenarios of U. S. Carbon Reductions: Potential Impacts of Energy-Efficient and Low-Carbon Technologies by 2010 and Beyond*. Oak Ridge and Berkeley: Oak Ridge National Laboratory and Lawrence Berkeley National Laboratory. ORNL-444 and LBNL-40533. September.

LINC. 1998. http://www.cpc.unc.edu/dataarch/nc/linc.html

Lindquist, P. 1997. Agricultural and GHG Emissions in the Northwest Ohio Global Change and Local Places Study Area. Paper presented at the Meetings of the East Lakes Division of the Association of American Geographers, Michigan State University, East Lansing, MI, October, 1997.

Lineback, N. G., T. Dellinger, L. F. Shienvold, B. Witcher, A. Reynolds, and L. E. Brown. 1999. Industrial Greenhouse Gas Emissions: Does CO_2 From Combusting of Biomass Residue Really Matter? *Climate Research*, **13**: 221–9.

Mayfield, M., B. Witcher, and T. Dellinger. 1996. Greenhouse Gas Emissions: Methane Released in North Carolina, 1990. *The North Carolina Geographer* **5**: 53–63.

Morioka, T., and N. Yoshida. 1995. Comparison of Carbon Dioxide Emission Patterns Due to Consumer's Expenditures in UK and Japan. *Journal of Global Environment Engineering*, **1**: 59–78.

Morioka, T., and N. Yoshida. 1999. Evaluation of Environmental Impact in Civil Infrastructure Systems with Respect to Sustainable Industrial Transformation. *Journal of Global Environmental Engineering*, **5**: 87–95.

Muraco, W. and S. Attoh. 1997a. Global Change and Local Places Northwest Ohio Study Area: Status Report. Paper presented at the Annual Meetings of the Association of American Geographers, Fort Worth, Texas.

Muraco, W. and S. Attoh. 1997b. Spatial Distribution and Trends in GHG Emissions in the Midwest. Paper presented at the Meetings of the East Lakes Division of the Association of American Geographers, Michigan State University, East Lansing, Michigan.

Reid, N. 1997. Economic Restructuring and GHG Emissions in the Northwest Ohio Global Change and Local Places Study Area. Paper presented at the Meetings of the East Lakes Division of the Association of American Geographers, Michigan State University, East Lansing, Michigan.

Stern, P. C., O. R. Young, and D. Druckman, eds. 1992. *Global Environmental Change: Understanding the Human Dimensions*. Washington, DC: National Academy Press.

United States Environmental Protection Agency. 1995a. *Inventory of U. S. Greenhouse Gas Emissions and Sinks: 1990–1994.* Washington, DC: U. S. Environmental Protection Agency, Office of Policy, Planning, and Evaluation. EPA-230-R-76-006.

United States Environmental Protection Agency. 1995b. *State Workbook; Methodologies for Estimating Greenhouse Gas Emissions,* 2nd edition. Washington: U. S. Environmental Protection Agency, Office of Policy, Planning, and Evaluation, State and Local Outreach Program. EPA-230-B-95-001.

United States Environmental Protection Agency. 1996. AIRS Data Base. www.epa.gov/airs/aeusa/index.html

United States Environmental Protection Agency. 1997. Pennsylvania Greenhouse Gas Emissions and Action Summary. State and Local Outreach Program, US Environmental Protection Agency. http://134.67.55.16:7777/DC/GHG.NSF/ReportLookup/PA

United States Environmental Protection Agency. 2001. http://www.epa.gov/globalwarming/emissions/national/gwp.html

United States Environmental Protection Agency, Office of Transportation and Air Quality. 2002. *MOBILE5 Vehicle Emission Modeling Software.* http://www.epa.gov/oms/m5.htm

United States Global Change Research Program. 1994. *Our Changing Planet. The FY 1995 U. S. Global Change Research Program.* Washington: USGCRP.

Woods & Poole Economics. 1998. *County Projections to 2025: Complete Economic and Demographic Data and Projections, 1970 to 2025, for Every County, State, and Metropolitan Area in the U. S.* Washington, DC: Woods & Poole Economics. http://www.woodsandpoole.com/

Yarnal, B., and C. Polsky. 1998. *1990 Inventory of Greenhouse Gas Emissions for Centre, Clearfield, Clinton, Snyder and Union Counties, Pennsylvania.* Prepared by B. Yarnal, C. Polsky, J. Adegoke, M. Alcaraz, J. Carmichael, A. Denny, P. Mitchell, B. Reifenstahl, and D. Vorhees.

PART TWO

Learning from localities

3

Global change and Southwestern Kansas: local emissions and non-local determinants

John Harrington Jr., David E. Kromm, Lisa M. B. Harrington,
Douglas G. Goodin, and Stephen E. White

Landscape, life, and livelihood

The Southwestern Kansas study area (Figure 2.1) lies within the American High Plains, a semi-arid region west of the 100° meridian that extends northward from West Texas through Kansas and Nebraska to the Dakotas. Characterized by nineteenth century explorers as the 'Great American Desert,' the first European-American landholders introduced cattle ranching to this nearly treeless shortgrass prairie. Through the Homesteading Act and the efforts of railroads, much of the High Plains was settled by crop farmers in the late nineteenth century. The area of land successfully cultivated varied with precipitation cycles, however, and the particularly long dry spell that occurred in the 1930s resulted in land abandonment through much of the region, lending the High Plains a new name: The Dust Bowl.

The study site lies at the center of the High Plains and in the heart of the former Dust Bowl. Its six counties encompass an area of approximately 14,120 square kilometers (5,450 square miles), inhabited by slightly more than 90,000 people and over 900,000 cattle. The three principal settlements are Garden City (2000 population of 28,451), Dodge City (25,276), and Liberal (19,666). Southwestern Kansas lies at relatively high elevation: 610–1,070 m (2,000–3,500 ft) above sea level, but contains little internal topographical relief.

The study area's climate is semi-arid with mean monthly temperatures ranging from -1 to $+27\,°C$ (30–80 °F). Precipitation averages less than 58 cm (23 in) per year, most of which falls as spring and summer rain (Goodin *et al.* 1995). Variability and severity characterize weather throughout the High Plains and in the study area (Rosenberg 1986). Droughts occur periodically, as in the 1930s and the 1950s. High summer temperatures accompanied by strong winds are frequent (Lydolph and Williams 1982). Relatively low humidity generally provides moderate night-time temperatures in summer. Severe weather is common, especially in the spring and early summer, and winter can bring extremes. Winter snowfall is usually not excessive, but winds often exceed 60 kilometers per hour (37 miles per hour), creating blizzards that threaten livestock.

Irrigation with diverted river water began in the 1880s (Sherow 1990), but was limited by the capacity of the handful of streams to provide water. Widespread irrigation as a basis

The authors gratefully acknowledge the financial support provided by the United States Department of Energy's National Institute for Global Environment Change (NIGEC) Great Plains Center under award number LWT 62–123–065519.

Figure 3.1 Corn field irrigated by center pivot sprayer in the Southwestern Kansas study area.

for agricultural stability commenced with the tapping of the Ogallala Aquifer in the 1950s and 1960s. No longer a Dust Bowl, the Ogallala region of the High Plains became known as the 'New Corn Belt' or more poetically, the *Land of the Underground Rain* (Green 1973). New irrigation technology and the aquifer's abundant fossil water created a boom (Figure 3.1). The region's population grew and the agribusiness economy expanded rapidly in the 1970s and 1980s. Irrigated crop land in the study area increased from 178,000 hectares (440,000 acres) to over 346,000 hectares (855,000 acres) between 1969 and 1997 (United States Department of Agriculture, National Agricultural Statistics Service 1969, 1997). The regional economy has prospered on the basis of irrigated forage crops, which the area's farmers feed to livestock.

Much of the dominant cattle industry centers on huge feedlots whose cattle collectively consume many tons of grain daily (Figure 3.2). The cattle, in turn, flow to the four large packing plants that now operate within the study area. One, an Iowa Beef Packers (IBP) facility near Holcomb, KS, can process more than 6,200 animals a day. Through its multiplier effects, irrigated agriculture underlies all aspects of the regional economy (Kromm and White 1992). Can the current prosperity be sustained or will the aquifer be depleted sufficiently to lead to a bust? Deborah and Frank Popper forecast depopulation and a return to the dryland conditions of the early nineteenth century, suggesting that the region become part of a federally managed *Buffalo Commons* (Popper and Popper 1987).

Sustainability remains an open question. Where irrigation prevails, most rural towns seem vital and healthy. Where irrigation is absent, much less economic activity is evident (White 1994). Water conservation has become more prevalent and is seen as the answer to

Figure 3.2 Brookover Feedlot in the Southwestern Kansas study area.

sustaining irrigation and the integrated agribusiness economy it supports. Municipalities, factories, and other water users have decreased their water use, but irrigation remains a major concern. More than 90% of the water consumed on the High Plains is used to irrigate crops. If irrigators can substantially reduce water consumption, most will continue to enjoy adequate and accessible water for many decades to come. But while irrigation efficiency has improved significantly in many areas, the total number of irrigated hectares continues to expand.

Natural gas and petroleum are extracted in Southwestern Kansas from the Hugoton Gas Area. Production from the Hugoton began in the 1920s and peaked in the early 1970s. The field is now in the autumn of its life, but has recently seen an expansion of gas extraction owing to new rules allowing an additional well in each section (260 ha) of land.

The Southwestern Kansas study site was chosen in part because of the environmental and social changes that have occurred in response to the dynamic changes in the local integrated agribusiness economy. Expanded irrigation, confined cattle feeding, and beef packing have attracted a number of supporting industries, including package and container plants, fertilizer and agricultural chemical manufacturing, electrical power generation, and transportation (mostly trucking). Recent years have seen the expansion of corporate hog production and dairying as well. These changes have sparked a significant increase and redistribution of population. Farm consolidation and improved technology have decreased the demand for on-farm labor; hence, many rural areas are losing people who migrate to local urban centers. The demand for workers in factories and packing plants, however, has increased well beyond what the local labor pool could supply. As a result, substantial in-migration of people, many of Latino or Southeast Asian origin, has occurred. These population changes have introduced some distinct and significant problems, including such issues as provision of housing, social services, and cultural amenities for a rapidly growing population with new and dissimilar cultures (Christian 1998).

Greenhouse gas emissions: surprisingly diverse

Localities contain both natural and anthropogenic forces that alter local radiative forcing and potentially affect larger parts of the Earth system. Factors requiring assessment include changes in trace gas emissions and sinks, and the ways in which land-cover changes affect

---

Box 3.1 Water vapor

The recent transformation of the landscape in Southwestern Kansas has come about in large part because of the availability of groundwater for irrigation. Traveling through the area in the summer, one is struck by the number of hectares where water is being applied. While some of this water is stored in the plants to which it is applied, evaporation and plant transpiration deliver considerable amounts of water vapor to the atmosphere. Since water vapor is an important natural greenhouse gas, local changes in low-level atmospheric moisture content were assessed to determine whether the use of ground water was affecting local climate. Hourly observations of dew point temperature for Dodge City were obtained for the period 1948–95. These data, summarized as monthly means and temporal trends for the months of March, April, and May, suggest increases in near-surface atmospheric moisture beginning about 1970. In addition, the time series graphs for all three months suggest that year-to-year variability decreased in the second half of the time period (Harrington 1998).

Results from comparisons of dewpoint temperatures for the two halves of the time series indicate a significant increase from 1972 to 1995, with substantial increases for the spring months of March, April, and May. These increases are even greater for minimum (overnight low) dewpoint temperatures. For example, the average surface dewpoint temperature for March at Dodge City increased from $-4.2\,°C$ ($24.4\,°F$) for 1948–71, to $-1.7\,°C$ ($28.9\,°F$) for the 1972–95 period. The $2.5\,°C$ ($4.5\,°F$) increase in surface dewpoint temperature for March corresponds to a 22% increase in atmospheric water vapor pressure.

The area of irrigated winter wheat, corn, and sorghum in Ford County increased from fewer than 5,000 hectares (12,350 acres) in the 1960s to more than 25,000 hectares (62,000 acres) by the mid-1980s. The timing of the increase in irrigated area in Ford County corresponds with the upward trend in near-surface dewpoint temperature. Groundwater utilization may well be the cause of the increased low-level atmospheric moisture evident in the spring months.

Table 3.1 Global warming potential in Southwestern Kansas

(1,000 metric tons per year of equivalent carbon dioxide)

Rank	1990		1980		1970	
1	Utilities	2,224	Industry	2,343	Industry	3,312
2	Industry	2,103	Transportation	802	Transportation	546
3	Transportation	1,098	Domestic stock	801	Domestic stock	505
4	Domestic stock	864	Animal manure	699	Utilities	407
5	Animal manure	759	Utilities	288	Manure	393

absorption and reflection of solar and terrestrial radiation. Documenting relevant local changes in Southwestern Kansas began with the use of modified Environmental Protection Agency protocols to estimate greenhouse gas emissions from available local data. Emissions estimates were generated for three greenhouse gases (carbon dioxide, methane, and nitrous oxide) for the years 1990, 1980, and 1970. The magnitudes and the diversity of the emissions were unanticipated by the members of the Global Change and Local Places team, as was at least one of the sources (Box 3.1).

Goodin *et al.* (1998) document the rationale and methods used to estimate 1990 greenhouse gas emissions for Southwestern Kansas. Data for the major fossil-fuel-burning carbon dioxide emitters were obtained from the United States Environmental Protection Agency Aerometric Information Retrieval System (AIRS) self-reported data base. Since AIRS data were not available for 1980 or 1970, it was necessary to backcast several fossil fuel emissions categories by using surrogate data. The Kansas Global Change and Local Places team used the relationship between annual data on local natural gas production obtained from the Kansas Geological Survey and AIRS data for the period 1986–1996 (Holden *et al.* 1997).

The global warming potential from emissions in the six-county study area is estimated at 7,615,000 metric (8,390,000 short) tons of carbon dioxide equivalent for 1990. Greenhouse gas emissions estimates were lower in both 1980 at 5,477,000 metric (6,035,000 short) tons of carbon dioxide equivalent and 1970, when an estimated 5,732,000 metric (6,318,000 short) tons of carbon dioxide equivalent were emitted. The 1980 estimate is the lowest of the three because natural gas production and transmission through interstate pipelines were significantly lower during this period. A major reason for increased emissions in 1990 was the addition of a new coal-fired electric power plant that went on line in 1983. Some variation exists in the lists of the five largest emitter categories for each of the three years (Table 3.1). These five categories accounted for more than 90% of all greenhouse gas emissions in 1970, 1980, and 1990.

Because almost all readily visible landscape changes (irrigated crops and feedlots) in Southwestern Kansas are associated with animal agriculture, the Kansas team's working hypothesis was that greenhouse gas emissions (primarily methane and nitrous oxide) from the large numbers of livestock and from the fertilizers used on feed grains would comprise a larger percentage of emissions for the study area than for Kansas or for

Box 3.2 Land-cover change

Landsat Multispectral Sensor data were used to assess local changes in surface reflectivity and vegetation greenness (Goodin *et al.* 2002). Three imagery dates in spring, summer, and fall were obtained for each of five years (1972–73, 1975–76, 1983, 1987, and 1992). Limited cloud-free imagery necessitated selecting from two consecutive years for the first two periods. The Landsat Multispectral Sensor data were georectified to Universal Transverse Mercator coordinates (Jensen 1996) and converted from raw digital values to reflectance using Robinove's (1982) method. In order to evaluate vegetation greenness trends, reflectance data were converted to the Normalized Difference Vegetation Index. In addition, the three dates for each year were combined into one large 12-band data set for use in land-cover classification.

Unsupervised classification of the imagery was done by using the ISODATA algorithm (Richards 1993) and four broad land-cover classes: urban–water, rangeland–pasture, cool-season crops, and warm-season crops. The resulting maps were used to evaluate land-cover change in the study area over the twenty years from 1972 to 1992. Results suggest that the majority of land-cover change between 1972 and 1992 occurred as a result of agricultural land-use expansion and intensification. In 1972–73, about 40% of the land was cultivated. By 1992, the proportion had increased to 61%. Much of the increase consisted of land irrigated by center pivot systems. Dominance by cropping system (warm-versus cool-season crops) showed an unexpectedly dynamic pattern. The local demand for feed grains (mostly warm-season crops) increased sharply from 1972 to 1992, but the overall proportion of warm-season crops did not. The total area devoted to warm-season crops did increase, however.

Mean reflectance values for the crop and pasture areas show no dramatic changes over the twenty-year period. Cultivated crops have the highest mean reflectance, with warm-season crops slightly brighter than cool-season land cover. Rangeland is least reflective. Overall, mean values for the three vegetated cover types show a slight upward trend over the five image dates, suggesting that, even though the reflectivity of individual land-cover types has not changed significantly, the overall reflectivity of the study area has increased. The explanation of this seeming anomaly appears to lie in the relative mix of cover types. Brighter, more reflective cultivated crops have replaced less reflective pasture land. Thus, increased reflectivity can be directly attributed to human-induced land-cover changes.

Vegetation greenness measured by Normalized Difference Vegetation Index values decreased slightly between 1972–73 and 1975–76, and then it steadily increased through 1992. Given the increase in irrigation in the study area since 1975–76 this is not surprising. Increased rainfall would tend to increase greenness, whereas an association of greenness with irrigation expansion would tend to indicate a human cause for increased greenness. To conclude, changes in reflectivity resulted from alteration of relative proportions of individual cover types, where cover-type reflectivity remained roughly constant throughout the study period. In the case of greenness, however, both the overall mean and individual cover-type greenness values increased between 1972–73 and 1992, suggesting that greater greenness is not due to proportional changes in land cover, but rather to intrinsic changes in vegetation vigor, possibly due to climatic conditions.

the United States. Once the numbers from the agricultural sector were put in perspective, however, the relative importance of several point sources associated with energy production became obvious. Prior to 1990, the industrial emissions category, which in Southwestern Kansas is dominated by the emissions from natural gas compressor engines, was by far the most important. When domestic stock and animal manure data are combined, the importance of animal agriculture emissions in the study area (21.3% in 1990) becomes more evident. Therefore the four sectors of utilities, natural gas (which dominates the industrial category), livestock, and transportation were selected for attention as the region's primary greenhouse gas emission sources (see Box 3.2 for the role of reflectance).

Greenhouse gas emissions drivers: the importance of national and global forces

The major sources of greenhouse gas emissions in Southwestern Kansas are diverse and subject to local and external influences. Structure, in the sense of external governmental regulations, greatly affects electrical power generation, which accounted for 29.2% of the study area's greenhouse gas emissions in 1990. The same holds true for natural gas extraction and transmission, which release most of the 27.6% of emissions accounted for by the industrial category. Distant corporate owners control the production and distribution of natural gas. Local residential demand for natural gas and electricity remains low because of the sparse population. Local agency prevails, however, in confined animal feeding (21.3% of study area emissions), in transportation (14.4%), and in power generation by the regional rural electric cooperatives that own and operate the single large facility. Most feedlots and transportation companies are owned and managed by investors and firms within the six-county study area. Local control is critical in understanding the driving forces underlying the major sources of greenhouse gas emissions for both confined feeding and transportation, whereas external control dominates natural gas. Power generation responds to both local and external influences.

Animal agriculture

Animals have been of central importance in the Southwestern Kansas study area since the earliest human occupation of the region. The transitions from hunting bison, to cattle ranching, and then to confined feeding of cattle have all occurred within the past 150 years. This tradition of animals and adaptability suggests that livestock will underpin the local economy for decades to come, and that greenhouse gas emissions from animal agriculture will continue to increase. Feedlots have been used for only 50 years in the area, and the proportion of cattle on feed in the study area increased from 56% of all cattle in 1974 to more than 72% in 1999, when the number exceeded one million head.

Physical and economic forces underlie the growth of confined cattle feeding. Fed cattle gain mass more rapidly and efficiently and they consistently yield higher-quality meat. The region offers almost ideal climatic conditions for cattle in the form of relatively low humidity and mild winters. Feed grains and groundwater are available in abundance, the four large beef packing plants in the region offer ready markets, and state and local environmental regulations are cattle-friendly, as is the study area's isolation from large population centers whose residents might object to some of the fragrant byproducts of livestock raising and processing. The entry of mega-dairies and confined hog feeding will result in additional greenhouse gas emissions. Recent trends and discussions with local decision makers suggest the study area will support more animals (and therefore generate more methane) in the next 20–25 years. Only genetic changes in the animals, different feed rations, or improvements in waste management will enable the region to avoid producing more animal-based greenhouse gas emissions in the future. Local forces play the dominant role in the confined feeding industry of Southwestern Kansas, but a decline in the demand for the Southwestern Kansas product, caused by a decrease in beef consumption or a significant increase in beef imports, would reduce cattle numbers.

Natural gas

The Hugoton Gas Area accounts for more than 90% of the natural gas produced in Kansas and 3.7% of the gas in the United States national market. The three westernmost counties of the study area partly overlay the Hugoton field and account for about 20% of all natural gas production in Kansas. In addition to local extraction, a considerable volume of gas flows through the study area and the state in interstate pipelines. Both the extraction of natural gas from underground reserves and transmission through pipelines to national markets produce greenhouse gas emissions (primarily carbon dioxide). Most of these carbon dioxide emissions originate in large compressor engines used to either pull the gas rapidly from underground strata or push the gas through the pipelines. Leaks of natural gas from old pipelines and accidents do occur. The volume of greenhouse gas emitted in this manner, while difficult to quantify, is assumed to be very small in comparison to compressor station emissions (Figure 3.3).

Natural gas was first discovered in the Hugoton area in 1922, and experts estimate that 65–75% of the gas has already been extracted. Recently, the number of wells allowed per section was doubled by state action in order to increase the efficiency of extracting the remaining resource. Wellhead pressures are declining. Initially pressures exceeding 30.5 kg cm^{-2} (435 lbf in^{-2}) in the 1930s have steadily dropped to a current average of only 5.3 kg cm^{-2} (75 lbf in^{-2}). Extraction will likely become more expensive in the future as additional compression is needed. The additional compression will produce

Figure 3.3 Natural gas compressor station in the Southwestern Kansas study area.

increases in greenhouse gas emissions per cubic foot of gas produced. As the Hugoton field plays out, however, overall declines in total greenhouse gas emissions from production are likely. Compression of natural gas flowing through Southwestern Kansas in interstate pipelines and the resulting greenhouse gas emissions are likely to fluctuate over the next 20–25 years as external markets seek gas from least-cost areas, including domestic and Mexican fields that would supply gas for transmission through Southwestern Kansas.

Utilities

The single largest source of greenhouse gas emissions in Southwestern Kansas is Holcomb Station, a coal-fired electric power-generating facility owned by the Sunflower Electric Power Cooperative, a non-profit, consumer-owned, generation and transmission utility. The Sunflower plant was built in the early 1980s. Holcomb Station is a 362 MW plant with a base demand of 200 MW (summer demand has reached as high as 295 MW). Unit trains bring low-sulfur coal to the plant from Wyoming. By 1990, utilities was the largest greenhouse gas emissions category, having exceeded the emissions produced by the industrial and gas production and transmission categories. External forces were dominant in determining that a coal-fired plant be built in an area where natural gas was abundant: the Power Plant and Industrial Fuels Use Act of 1978 stipulated that natural gas could not be used for base load production (Box 3.3).

Box 3.3 Southwestern Kansas coal-fired power plant: external control and inadvertent results

Electricity generation currently ranks as the most important source of carbon dioxide emissions and global warming potential in the Southwestern Kansas study area, a somewhat surprising finding given the relatively low population density of this largely rural and agricultural region. A coal-fired power plant, Holcomb Station, is the largest single source of emissions in the six-county study area. The construction of a power plant built to burn coal in a major natural-gas-producing region seems bizarre. Natural gas (methane) is more easily handled than coal, and is regarded as a cleaner fuel because of its lower sulfur content, and combustion generally emits much less carbon to the atmosphere than does burning coal (Pelham 1981; United States Department of Energy 1997b). Holcomb Station operates exclusively with low-sulfur coal transported by rail from the Powder River Basin in Wyoming. Southwestern Kansas' utility-related greenhouse gas emissions are an important illustration of how government policy designed to address certain national concerns may have unanticipated side effects.

Holcomb Station is the major generating facility of Sunflower Electric Power Corporation, a non-profit, cooperative rural electric generation and transmission utility. Sunflower is owned by six rural electric cooperatives, and provides power to seven electric cooperatives in 34 western Kansas counties. Excess production is sold to other electric power providers, resulting in reduced rates for cooperative members. Rising electricity demand led Sunflower to upgrade its Holcomb Station to increase production capacity in late 1997.

Oil and natural gas use in the United States increased from the 1950s into the 1970s (Pelham 1981). Shortages and higher petroleum and natural gas prices in the early 1970s, largely related to the Arab oil embargo (United States Department of Energy 1997a), led political leaders to search for policies that would reduce dependence on imported fossil fuels and conserve domestic petroleum and natural gas. Switching to nationally plentiful coal was seen as a means to these ends. Because of technological difficulties in converting other fuel users away from oil and gas, the one industry still fairly heavily dependent on coal – electric power generation – was targeted by the United States Congress (Pelham 1981).

Planning for the Holcomb Station came at a time when the nation was highly sensitive to energy issues. Construction of the station was approved in 1978, and the facility was built by Sunflower 'to ensure a reliable source of electric power for western Kansas' (Harrington and Kaktins 1998). The design and construction of this coal-fired power plant was affected by federal policy responses to energy concerns, including the Energy Supply and Environmental Coordination Act of 1974 (PL 93-319) and the Powerplant and Industrial Fuels Use Act of 1978 (PL 95-620). The Energy Supply and Environmental Coordination Act (ESECA) allowed the federal government to prohibit the use of natural gas and petroleum by electric utilities (Buck 1978; United States

Department of Energy 1997a). The Powerplant and Industrial Fuels Use Act (FUA) succeeded ESECA and extended federal powers regarding oil and gas use (United States Department of Energy 1997a) by specifying that no new base load facilities could rely on coal.

In the 1980s, growth in federal fuel use restrictions slowed. The Omnibus Budget Reconciliation Act of 1981 (PL 97-35) reduced federal authority to issue oil and natural gas use prohibitions (United States Department of Energy 1997a). In response to the recognition that natural gas was more plentiful than had been assumed, FUA was amended to repeal the provision that all electrical utilities convert from natural gas by 1990. In 1987, another amendment to FUA rescinded its prohibitions on the use of oil and gas in new power plants and on construction of power plants able to use an alternate fuel. The new provision mandated that all new base load power plants designed for oil or gas be convertible to coal or an alternate fuel (Schorr 1992). These revisions came too late for Holcomb Station, however.

Political decisions often have unforeseen side effects. When legislation is focused on a single hot issue such as a sudden desire to conserve natural gas, the results may lead to surprising or undesirable outcomes. The end result of FUA was to establish a coal-fired power plant in a natural-gas-producing area, thereby also creating higher greenhouse gas emissions. The electric utility industry currently is in a great state of flux, as a consequence of federal legislation in the form of the Energy Policy Act of 1992 (PL 102-486) designed to create greater competition. Some in the industry wonder whether the law actually will benefit rural electric customers or will favor urban areas and quantity consumers of electricity. Restructuring could cause a decline in state and electric utility demand side management programs that promote energy efficiency (Sissine 1998). However, energy supply and cost problems arising (especially in California) in 2000–01 have brought the form and extent of further restructuring into question.

The future of greenhouse gas emissions from electric power generation in the Southwestern Kansas study area and elsewhere is difficult to forecast because of questions regarding power industry restructuring and potential actions to reduce emissions at the national and international levels. International greenhouse gas emission reduction agreements and domestic policy changes may eventually affect electric power generation and associated greenhouse gas emissions in the study area. Many electric utility industry officials are highly suspicious of emissions reduction efforts, but the industry responds to federal directives as necessary.

External forces in the form of federal laws and regulations have shaped the character of electric utilities and related greenhouse gas emissions in Southwestern Kansas. External influences will continue to play a role in determining economic and environmental conditions for such localities, and those influences must be considered when addressing wide reaching environmental, social, and economic trends (Harrington and Kaktins 1998).

Uncertainty marks the future of electric power generation in Southwestern Kansas, and external forces will play a major role in determining local outcomes. Competition in the electric power industry resulting from the Energy Policy Act of 1992 and Federal Energy Regulatory Commission Orders 888 and 889 may lead to a form of deregulation called retail wheeling (Harrington and Kaktins 1998). Since Holcomb Station is a recently constructed facility, the debts incurred in construction are costs to the utility that are not recoverable at market-based rates. These stranded costs ('stranded assets'), call into question the viability of Holcomb Station and Sunflower Electric Power Corporation. Future utility greenhouse gas emissions in Southwestern Kansas are uncertain, with changes likely to be the result of decisions made external to the study area. In spite of such questions, Sunflower Electric Power Cooperative is recommissioning an older gas-fired power plant and has recently upgraded the coal-fired plant to serve a new peak level. These changes reflect increases in demand, and new arrangements have been made to improve the vitality of the company.

Transportation

Transportation accounted for 14.4% of the Southwestern Kansas study area emissions in 1990 or 1.1 million metric (1,212,200 short) tons per year of carbon dioxide equivalent. County Business Patterns data from 1965 through 1995 strongly support the proposition that a significant component of the growth in the transportation sector is attributable to trucking. Both the number of trucking establishments and persons employed in trucking have more than doubled in the region in the past thirty years. In addition to increased use of large trucks to supply goods to the relatively small residential sector, the trucking industry in Southwestern Kansas moves animals and feeds among farms, feedlots, packing plants, and markets. Trucks connect the fertilizer and fuel sources with the farms, the feed grains with the feedlots, and the feedlots with their cattle suppliers and the meat packing plants. A smaller proportion of the transportation emissions are accounted for by shipments of boxed beef and through truck traffic on the major highways that cross the study area.

Changes in truck traffic in the study area can be linked to the region's growing population, the rapid growth of feedlots, meat packing, and associated agricultural activities, and to population growth in the area's regional centers (Dodge City, Garden City, and Liberal). Local investments in trucking-related business (sales of trucks, refrigerated trailer construction, service facilities) continue, and trends suggest increased truck traffic and associated emissions in the future. Although transportation responds to regional demand, local needs to move large quantities of materials (whether it be feed grain, cattle, boxed beef, or manure) will keep the transportation sector growing.

Implications

The diversity of greenhouse gas emission sources in Southwestern Kansas, their relative consistency through time, and the balance between local and external control, suggest that future emissions in the area will not differ fundamentally from those of the past, and that the study area has some influence over its own destiny as an emissions source. Utilities, natural gas, confined animal feeding, and transportation accounted for 90.1% of all greenhouse gas emissions in 1970 and 1980 and for 92.5% in 1990. Natural gas will be a declining source of emissions as extraction wanes, but interstate gas transmission pipelines will keep the industry in the mix. Increasing demand for electricity should sustain power generation as a source of greenhouse gas emissions unless the Holcomb station is shut down as uneconomic and the area is then served by a distant plant.

Animal agriculture and transportation are closely connected to each other and to the irrigated cultivation of feed grains and the large-scale processing of beef in study site packing plants. If these activities remain competitive in the marketplace, feedlots and trucking will continue to be major sources of greenhouse gas emissions. A recent trend that could increase future methane emissions is the movement of dairies and hog operations into the study area.

Reducing greenhouse gas emissions: limits to local agency

The greenhouse gas emissions setting for Southwestern Kansas is in some ways more complex than for the other Global Change and Local Places study areas. Several major source activities contribute similar proportions of the total quantity of carbon dioxide equivalent forcing, and to a large extent these activities are driven by external forces. Global economic relations (particularly in food demand and production), policy at the state and national levels, and corporate energy decisions made at national or international headquarters outside the region all influence emissions from this area. That is not to say that reductions in emissions are entirely beyond the control of the local area: reductions could be made, although probably not with the local level of control that might be exerted in some other regions of the country.

Local managers of specific sources can take actions to reduce emissions to some extent. The greatest local control exists in the feedlot and transportation sectors. Owing to the strong influence of national policy and regional to international energy economics on both the electric and natural gas industries, local decision-making is much less likely to result in local emissions reductions by these source sectors. Residents of an area always have some level of control over local greenhouse gas emissions. There are a number of ways in which they affect emissions, including the level of demand for residential energy, specifically that based on fossil fuel combustion, and their willingness to promote or support both energy efficiency measures and emissions reduction measures by local businesses. The demand for energy can be affected by local willingness to adopt such alternative energy sources as solar and wind power, willingness to adopt energy-efficient practices and technologies at home, and willingness to reduce vehicle fuel consumption. Because local residents are employees, customers, and operators of local emissions sources (i.e. feedlots, electric power cooperatives, natural gas companies, and transportation enterprises), attitudes toward greenhouse gas emissions may create an environment where more actions to reduce emissions are taken (or not).

Local attitudes

Two methods were used to gather information regarding local knowledge and attitudes toward greenhouse gases (Harrington 2001; Harrington & Lu 2002). First, interviews were conducted with a limited number of individuals in leadership positions and in the most important greenhouse gas source sectors. Second, a household questionnaire was used to gain a broad perspective on the perceptions of Southwestern Kansas residents. A total of 702 surveys were mailed to randomly selected households in the six-county area; of the 584 deliverable questionnaires, 199 usable responses were received, yielding a usable response rate of 34%.

Interviews

Interviews with local leaders, either in specific areas of economic activity or in government positions, were conducted to obtain information about the outlook or attitudes of those in the best positions to pursue change in the local area. Field and telephone interviews dealing with greenhouse gas knowledge and potential were conducted in August and November 1998. The interviewees were specifically selected for the Global Change and Local Places project. Twenty sources supplied information regarding climate change and mitigation attitudes. The interviews were unstructured, relatively informal discussions, but they addressed four general topics: level of concern, mitigation strategies, desirable (societal) levels of emissions control, and technology awareness.

The attitudes of those interviewed and their impressions of the views held by other local residents reveal that:

- The reality of global climate change is credible, but it is also highly variable. To some, it is a real concern; to others, it is a subject of interest, and to still others, it is an unproven conjecture and consequently of no concern.
- Most of those interviewed stress the need for *proof* of the reality of climate change before they will accord the matter serious attention.
- Climate change is therefore *not* a major current concern in the area.
- Some of the concern that does exist arises from fears that greater regulation of greenhouse gas emissions will harm specific industries without addressing a real problem.
- Those interviewed *do not connect* actions in the study area to conditions in other places.
- Specifically, respondents don't connect study area greenhouse gas emissions to enhanced greenhouse effects. Energy savings were not necessarily seen as related to reducing greenhouse gas emissions.
- Carbon dioxide was mentioned as possibly benefitting crop growth by some interviewees.

Strategies that could decrease greenhouse gas emissions may be in place in response to other problems, but no strategies have been specifically undertaken in response to concerns about climate change. The best selling point for any mitigation action is likely to be its economic benefits. Public relations may also help promote mitigation, though some respondents saw themselves in a powerless position, viewing global warming as 'a problem farmers can't do anything about.' An interviewee employed by the electric power industry contended that conservation is *not* a strategy.

Local residents desire goals set locally or at some middle level of decision-making. A need for some federal or state oversight is recognized, but the level of government that will actually be trusted lies closer to home. The federal government is viewed as capricious in policy development and implementation and as inconsistent in its year-to-year legislation. Residents were especially wary of unfunded mandates: goals or regulations issued from on high with no support for their realization. In spite of these general attitudes, local control sometimes is seen as too pliable; the potential for local self-interests to control actions is perceived by some as a potential problem. Cooperation between locals, trade organizations, and government agencies is preferred to the imposition of regulations. In some industries, state- and federal-level control is preferable, partly as a matter of convenience or familiarity. For the natural gas industry, current control is at the state and federal levels. Some feedlot-associated respondents also favored state control. The belief that people will act on their own when convinced a problem exists was mentioned repeatedly.

Knowledge of specific technologies varied. Interviewees who are deeply involved in specific enterprises, such as transportation or feedlots, often appear to be well informed regarding available technologies – if not technologies for greenhouse gas management, then practices resulting in greater efficiency. Efficiency-oriented techniques often yield emissions reduction co-benefits. Respondents also recognize the need for outside expertise, demonstration projects, and joint ventures that involve governments and educational institutions.

Household questionnaires

A broad range of opinions regarding climate change exists among Southwestern Kansas households. More than a third of respondents said their level of concern about the effects of greenhouse gas emissions was low or very low. A smaller proportion (14.6%) thought there would be no problem at all from climate changes caused by global warming over the next 50–100 years if no action is taken. Conversely, 26.1% said such climate changes will be very serious or extremely serious if no action is taken. Several respondents expressed great distrust of government. One opined that the idea that burning oil, coal, and natural gas for energy will cause global warming in the future is 'government propaganda.' More generally, 28.6% considered global warming mostly an unproven theory, whereas 21.1% said it was mostly a proven fact, and 30.7% considered the forecasts of global warming somewhere between unproven speculation and proven fact. One third of household respondents thought themselves somewhat uninformed or completely uninformed regarding climate change. Some respondents explained their lack of faith in suggestions that global warming be addressed by citing a need for more information:

- 'To think I could affect creation is crazy.'
- 'Not enough of time span or research years to be proven fact.'
- 'Need more proof of climate change. The 1970s were colder than normal, and people thought we were going to have an ice age. Now it's warmer than normal.'
- 'We are uneducated in global warming, but would be interested in learning.'
- 'More public awareness of ameliorating measures needs to be promoted and information provided.'

There is also a significant component of the population with sincere environmental concerns and with a deep feeling of responsibility. A number of residents argued that global warming must be addressed, that individuals have a responsibility to do so, and that cooperation is needed. However, climate change is more nebulous than other environmental issues, being viewed as less conclusively proven, not a local problem, less immediate than other problems, or better addressed at other locations.

Overall assessment of attitudes and implications

The viewpoints expressed by leaders and the comments made by household respondents are consistent. Climate change is not seen as a major concern in the study area by either group. Both exhibited considerable internal variation with respect to level of concern over climate change and the credibility of that threat. Both groups wanted proof that a problem exists prior to taking action. Both groups had difficulty connecting Southwestern Kansas to any climate or greenhouse gas problem that may exist elsewhere or at broader scales. Paradoxically, many respondents also feel responsible for the environment.

While some knowledge of technologies or actions available for decreasing greenhouse gas emissions is extant among both groups, it is not connected to specific strategies for emissions reductions. Actions that have been implemented were taken to meet other goals, most commonly economic or efficiency targets. A few respondents link greater energy efficiency with emissions reductions, but the link does not appear to be recognized when individuals take action to increase energy efficiency. Only 13% of household respondents reported having acted to reduce greenhouse gas emissions. However, when asked about particular energy efficiency strategies, a much higher proportion of respondents replied that such actions already had been taken. Other responses indicate a willingness to increase energy-use efficiency and thereby decrease greenhouse gas emissions. For most listed options, more residents were willing to adopt energy efficiency measures than were unwilling, though reductions in driving and adopting solar energy were notable unpopular exceptions.

Federal agencies would not be trusted (perhaps to the point of conscious resistance) to manage attempts to reduce or control greenhouse gas emissions. The federal level possibly could play a role as a generalist in such attempts, exercising broad oversight and coordination. Local or regional control is trusted more and seen as a more reasonable approach to specific local conditions. In spite of these expressed attitudes, responses to the questionnaire item asking residents to rank the level of responsibility for greenhouse reductions revealed that local inhabitants hold both individual households *and* federal government 'most responsible' for doing something about the problem. Local and state governments were seen as levels that should be held less responsible. Local self-interest is considered one barrier to effective action at that scale; lack of expertise is another major concern. Inhabitants view state universities and associated extension activities positively and as perhaps critical in educating and promoting appropriate methods for mitigating greenhouse gas emissions.

Overall, the opinions of Southwestern Kansans about greenhouse gases and global warming vary widely among the general populace as well as among community and emissions sector leaders. For *local* actions to be taken, more convincing evidence must connect climate changes and human-induced emissions for most residents to deem the problem credible and local contributions to its solution feasible. That said, a significant number of people expressed sincere concerns about environmental degradation and a willingness to take responsibility for arresting it. Marshaling opinion and action appear to depend on education and convincing proof that a threat exists and that local responses will yield positive effects. The media and state universities (particularly local extension service representatives) appear to possess the greatest capacities for affecting local opinion and promoting local mitigation.

Prospects are dominated by uncertainties

Whether Southwestern Kansas residents will develop the ability and the resolve to mitigate greenhouse gas emissions remains uncertain. The Intergovernmental Panel on Climate Change (Houghton *et al.* 1990) has generated several scenarios that estimate global greenhouse gas emissions for 2000, 2010, and 2020. These econometric models suggest local growth in greenhouse gas emissions. Technological developments, such as more efficient fuel use by large trucks or policies that tax carbon emissions, would doubtless reduce greenhouse gas outputs, but other factors suggest that local capabilities to change greenhouse gas emissions may be limited. A key consideration in assessing the capacity and propensity to mitigate is the local history of adaptation to earlier environmental changes (Easterling *et al.* 1993). Chapter 10 examines local responses to groundwater depletion in the Ogallala Aquifer as an analog to possible future responses to global climate change in the context of similar analogs from the other three study areas. A facet of local geography that may dampen local enthusiasm for mitigation is the sparseness of population in comparison to the large and growing livestock population in the study area. The current ratio of domestic animals to people exceeds ten to one. The dry climate, vast open spaces that provide a sense of isolation, a local population that cheerfully designates the odors from livestock feeding operations 'the smell of money,' easy access to major packing plants, and non-oppressive environmental regulations, combine to favor increased livestock numbers and greenhouse gas emissions. Moreover, a local knowledge base that connects improved energy efficiency with reductions in greenhouse gas emissions remains to be built. Even in the presence of the local desire to 'do good by the environment,' a substantial citizen education effort is prerequisite to energetic local action in the absence of some unforeseen catastrophe.

Greenhouse gas emission estimates for Southwestern Kansas for the years 2000, 2010, and 2020 were calculated from the baseline greenhouse gas emissions inventory for 1990. Estimated changes in social and economic conditions were extracted from data obtained from a commercial forecasting firm (Woods & Poole Economics 1998). Growth estimates used for projections include manufacturing employment, manufacturing earnings, farm employment, farm earnings, per capita income, housing units, population, and other factors related to economic growth (Dellinger 1997). In order to address the uncertainties inherent in forecasting, three separate scenarios were constructed to provide a range of possible emission levels. The medium projection follows the basic parameters of the Intergovernmental Panel on Climate Change Business-as-Usual Scenario, assuming little change in energy sources or emission controls (Houghton *et al.* 1990). In order to provide plausible

Table 3.2 Estimated rates of percentage growth in study area greenhouse gas emissions

Region	1990–2000	2000–2010	2010–2020	2000–2020	1990–2020
Southwestern Kansas	9%	12%	7%	20%	31%
Kansas	12%	11%	11%	23%	38%
United States	16%	14%	13%	30%	51%

Table 3.3 Southwestern Kansas study area emissions estimates

Figures are given in GWP $\times 10^6$.

	1990	2000	2010	2020
Low	7.62	6.92	7.72	8.36
Medium	7.62	8.46	9.30	9.98
High	7.62	10.38	11.58	12.53

emissions scenarios, a low projection (20% lower than the medium values) and a high projection (20% above the medium estimates) also are provided. Emission categories identified in the 1990 Global Change and Local Places emissions inventory are projected through the year 2020 using methods initially developed for the Global Change and Local Places Northwestern North Carolina study area (Dellinger 1997). The emissions projections are given in units of global warming potential based on a hundred-year time scale (United States Environmental Protection Agency 1995).

To evaluate the projections for the study area, future emissions estimates and estimated rates of increase were compared to those calculated for both Kansas and the United States. Baseline figures for the state and national estimates come from the United States Environmental Protection Agency Global Warming website (United States Environmental Protection Agency 1998) and the *U.S. National Greenhouse Gas Inventory* (United States Environmental Protection Agency 1995). The emission scenarios result in increased emissions except for a minimal decrease from 1990 to 2000 in the low projection scenario, though the rate of increase in Southwestern Kansas will be lower than the rates for Kansas or the United States (Table 3.2). Study area volatility differs from state and national variability also. In the medium projection, both Kansas and the United States display a downward trend in estimated rates of increase, but the study area rate increases from 9% for the period 1990–2000 to 12% for the period 2000–2010, and decreases to 7% for the period 2010–2020 (Table 3.2). These local fluctuations reflect shifts in estimated employment and earnings. Aside from the initial dip in estimated emissions for the low projection, all study area projections display a steady rate of increase in total emissions estimates throughout the period (Table 3.3).

Comparison of the low, medium, and high projections among individual counties in the study area suggests even greater local variability. Finney County was the dominant emitter of greenhouse gases in 1990 and all projections maintain this relationship. Greenhouse gas emission estimates for Finney County are more than double those of any other

Table 3.4 Southwestern Kansas greenhouse gas emissions by county, 1990

Figures are metric tons of equivalent $CO_2 \times 10^6$.

County	Utilities	Industrial	Animal Agr.	Transportation	Total
Finney	1.96	0.51	0.45	0.30	3.22
Ford	0.26	0.33	0.32	0.31	1.22
Gray	0.00	0.17	0.36	0.10	0.63
Haskell	0.00	0.24	0.32	0.08	0.64
Meade	0.00	0.19	0.14	0.08	0.41
Seward	0.00	0.68	0.22	0.22	1.12

county in the study area. The major difference between the more than 10,000,000 metric (11,020,000 short) tons of global warming potential for Finney County (high projection) and the other five counties is explained by the presence of the Holcomb Station power plant (Goodin *et al.* 1998; Harrington and Kaktins 1998). The 1990 estimates suggest that this single source is responsible for 1,960,000 metric (2,116,000 short) tons of global warming potential, or 26% of all study area emissions (Table 3.4).

Econometric projections suggest that greenhouse gas emissions in Southwestern Kansas are likely to increase. Yet two analog studies suggest that study area residents can adapt to environmental changes when motivated to do so. Southwestern Kansas lies in the heart of the 1930s Dust Bowl region that Easterling *et al.* (1993) examined as a part of the MINK (**M**issouri, **I**owa, **N**ebraska, and **K**ansas) study of whether modern farm practices and technology would enable farmers to cope with a return of drought conditions similar those of the 1930s. One of the MINK project's conclusions was that animal production in the Missouri–Iowa–Nebraska–Kansas area would be little affected by the return of such an analog climate. Cutter and colleagues (see Chapter 9) examined local responses to declining water levels in the Ogallala Aquifer as indicators of how Southwestern Kansas residents might respond to the changes that would accompany global warming. Past responses to that threat to local agriculture suggest that the study area's inhabitants are indeed likely to adapt to future environmental changes.

Local abilities to reduce greenhouse gas emissions will depend on both individual and corporate decisions to modify existing practices. Relevant examples for Southwestern Kansas include:

- change the feed mix for feedlot cattle (or other animals) to maintain growth rates and accomplish a reduction in animal-produced methane emissions,
- switch fuel at the Holcomb Station power plant to permit the use of local natural gas rather than Wyoming coal;
- shut down less efficient natural gas wells (those with low wellhead pressures); and
- use more fuel-efficient (lighter-weight or more aerodynamic) trucks.

All four actions will require investment and will encounter the normal human resistance to change. Global Change and Local Places staff interviews with local leaders and surveys of residents' beliefs regarding global warming suggest that currently the greenhouse gas reduction issue ranks low among immediate priorities, sometimes to the point of total uninterest. An intensive education effort would be needed to convince the local population that mitigating greenhouse gas emissions is worthwhile. Beyond local interest and motivation, many of the decisions that could reduce emissions must be made in corporate offices far distant from the study area.

One way for the Southwestern Kansas study area to accomplish an 8% or a 20% reduction in greenhouse gas emissions would be for the Holcomb Station power plant to shut down. A number of uncertainties cloud the future of the power plant (Harrington and Kaktins 1998), including the debt incurred to build the facility. One scenario involves *retail wheeling*, which permits individual consumers to choose a company from which to purchase electrical power. Power companies with large debts for recent construction may not be able to compete well in an open market in which formerly captive customers can buy power from one of several suppliers. Nevertheless the Sunflower Cooperative has shown an inclination to expand capacity and to enter into partnerships with other electric power companies, implying that its leaders are optimistic about its prospects.

Another matter to consider in thinking about future greenhouse gas emissions in Southwestern Kansas is the region's connections to its external markets. If significant fuel switching occurred that favored more natural gas use downstream on the interstate natural gas pipeline network, local greenhouse gas emissions might increase in Southwestern Kansas at the same time they decrease in more populous areas to the north and east. Similarly, if more animal producers shift to Southwestern Kansas, animal emissions will increase. An increase in manure application to agricultural land in Southwestern Kansas instead of processing it in lagoons elsewhere would decrease emissions, as spreading the waste releases smaller quantities of greenhouse gases.

In a national context, the Southwestern Kansas study area emits comparatively low quantities of greenhouse gases. Greenhouse gas emissions average 543 metric tons of carbon dioxide equivalent per square kilometer (1,535 short tons per square mile). The study area emits high per capita quantities of greenhouse gases (82.1 tons of carbon dioxide equivalent per person), but that number reflects both the sparse local population and the dominance of beef, electricity, and natural gas production for export to other regions. The forces driving change are also largely external to the study area: consumer demand for its products and corporate and federal decisions. Local attitudes toward climate change, anthropogenic greenhouse gas emissions, and programs to mitigate greenhouse gas emissions are currently ambivalent. With continued education, however, local support for ameliorative and adaptive programs would most likely increase. In the short run, Southwestern Kansas residents are more prone to accept adaptive measures than to engage in mitigation.

REFERENCES

Buck, E. 1978. Energy Users and Government Regulations. In L. Buck, ed. *Proceedings of the Third Annual Energy Users Law Seminar*, January 26–27. Washington: Government Institutes, Inc.

Christian, S. 1998. Latin Immigrants Fuel Dodge City's Meat-Packing Boom. *New York Times*, January 29.

Dellinger, T. 1997. *Global Change in Local Places: Blue Ridge/Piedmont Study Area Final Report.* Boone, NC: Appalachian State University Department of Geography and Planning.

Easterling, W. E., P. Crosson, N. J. Rosenberg, L. A. Katz, and K. M. Lemon. 1993. Agricultural Impacts and Responses to Climate Change in the Missouri-Iowa-Nebraska-Kansas (MINK) Region. *Climatic Change*, **24**: 23–61.

Goodin, D. G., J. E. Mitchell, M. C. Knapp, and R. E. Bivens. 1995. *Climate and Weather Atlas of Kansas: An Introduction*. Lawrence: Kansas Geological Survey. Educational Series 12.

Goodin, D. G., J. A. Harrington, Jr., G. I. Holden, Jr., and B. D. Witcher. 1998. Local Greenhouse Gas Emissions in Southwestern Kansas. *Great Plains Research*, **8**: 231–53.

Goodin, D. G., J. A. Harrington, Jr., and B.C. Rundquist. 2002. Land cover change and Associated Trends in Surface Reflecting and Vegetation Index in Southwest Kansas: 1972–1992. *Geocarto International*, **17** (1): 43–50.

Green, D. 1973. *Land of the Underground Rain: Irrigation on the Texas High Plains, 1910–1970.* Austin: University of Texas Press.

Harrington, J. Jr. 1998. A Climatology of Changing Spring Dew Point Temperatures: Dodge City, Kansas. *Preprints Ninth Symposium on Global Change Studies*: 114–16. Boston: American Meteorological Society.

Harrington, L. 2001. Attitudes Toward Climate Change: Major Emitters in Southwestern Kansas. *Climate Research* **16** (2): 113–22.

Harrington, L., and S. Kaktins. 1998. Policy and Local Utility Greenhouse Gas Emissions, Or: Why a Coal-fired Power Plant in a Natural Gas Production Area? *Papers and Proceedings of the Applied Geography Conferences*, **21**: 1–9.

Harrington, L. M. B., and M. Lu. 2002. Beef feedlots in Southwestern Kansas: Local change, perceptions, and the global change concept. *Global Environmental Change*, **12** (4): 273–82.

Holden, G., J. Harrington, M. Goodin, and J. DeHart. 1997. Estimating Changes in Greenhouse Gas Emissions for the Natural Gas Industry in Southwest Kansas. *Papers and Proceedings of the Applied Geography Conferences*, **20**: 20–4.

Houghton, J. T., G. J. Jenkins, and J. J. Ephraums, eds. 1990. *IPCC First Assessment Report. Scientific Assessment of Climate Change – Report of Working Group I.* Cambridge: Cambridge University Press.

Jensen, J. 1996. *Digital Image Processing: A Remote Sensing Perspective.* New York: Prentice Hall.

Kromm, D., and S. White. 1992. The High Plains Ogallala Region. In D. Kromm and S. White, eds. *Groundwater Exploitation in the High Plains*: 1–27. Lawrence: University Press of Kansas.

Lydolph, P. and T. Williams. 1982. The North American Sukhovey. *Annals of the Association of American Geographers*, **72**: 224–36.

Pelham, A., ed. 1981. *Energy Policy.* 2nd edition. Washington: Congressional Quarterly, Inc.

Popper, D. E., and F. J. Popper. 1987. The Great Plains: From Dust to Dust: A Daring Proposal for Dealing with an Inevitable Disaster. *Planning*, **53** (6): 12–18.

Richards, J. 1993. *Remote Sensing Digital Images Analysis: An Introduction.* Berlin: Springer-Verlag.

Robinove, C. J. 1982. Computation with Physical Values from Landsat Digital Data. *Photogrammetric Engineering and Remote Sensing*, **48** (5): 781–4.

Rosenberg, N. J. 1986. Climate of the Great Plains Region of the United States. *Great Plains Quarterly*, **7**: 22–32.

Schorr, M. M. 1992. *Legislation and Regulations Affecting Power Generation Systems.* Schenectady, NY: General Electric Company.

Sherow, J. 1990. *Watering the Valley.* Lawrence: University of Kansas Press.

Sissine, F. 1998. Energy Efficiency: Key to Sustainable Energy Use. Congressional Research Service Issue Brief for Congress. 4 March. http://www.cnie.org.nle/eng-16.html

United States Department of Agriculture, National Agricultural Statistics Service.1969. *1969 Census of Agriculture.* Washington, DC: USDA.

United States Department of Agriculture, National Agricultural Statistics Service. 1997. *1997 Census of Agriculture.* Washington, DC: USDA.

United States Department of Energy, Energy Information Administration. 1997a. *The Changing Structure of the Electric Power Industry: An Update.* http://www.eia.doe.gov/cneaf/electricity/chg_str/.

United States Department of Energy, National Laboratory Directors. 1997b. *Technology Opportunities to Reduce U. S. Greenhouse Gas Emissions.* October. http://www.ornl.gov/climate_change.

United States Environmental Protection Agency. 1995. *State Workbook; Methodologies for Estimating Greenhouse Gas Emissions,* 2nd edition. Washington: U. S. Environmental Protection Agency, Office of Policy, Planning, and Evaluation, State and Local Outreach Program. EPA-230-B-92-002.

United States Environmental Protection Agency. 1998. *Inventory of U. S. Greenhouse Gas Emissions and Sinks: 1990–1996.* Washington, DC: United States Environmental Protection Agency, Office of Policy, Planning, and Evaluation. EPA 236.

White, S. 1994. Ogallala Oases. *Annals of the Association of American Geographers*, **84**: 29–45.

Woods & Poole Economics. 1998. *County Projections to 2025: Complete Economic and Demographic Data and Projections, 1970 to 2025, for Every County, State, and Metropolitan Area in the U. S.* Washington DC: Woods & Poole Economics. http://www.woodsandpoole.com/

4

Northwestern North Carolina: local diversity in an era of change

Neal G. Lineback, Michael W. Mayfield, and Jennifer DeHart

Northwestern North Carolina offers a distinctive combination of economic development characteristic of the New South in the United States, extensive forest management, and intensive tourism. For project purposes, the region afforded opportunities to examine the effects of the population and affluence dimensions of the the $I = PAT+$ formulation, to consider the relative importance of forest carbon sinks in locality contributions to greenhouse gas emissions abatement, and to consider differences between local and statewide perspectives on greenhouse gas emission questions.

Greenhouse gas emissions in the Northwestern North Carolina study area originate primarily from a relatively small number of utility and industrial point sources, but affluence-related emissions from the transportation and residential sectors are also notable components of the region's mix of emissions sources. A surprisingly heavy reliance on local biomass waste by small manufacturing plants substitutes for the use of fossil fuels in the region, a facet of the global array of fuel and emissions issues perhaps evident in other forested regions.

Landscape, life, and livelihood

The study area covers parts of the Piedmont Plateau and Blue Ridge mountains with approximately two thirds lying in the Piedmont (Figure 2.2), a region characterized by rolling hills and gently sloping interfluves, with local relief typically less than 50 m (165 ft). The Blue Ridge portion of the study area is a heavily dissected mountainous region with most elevations lying between 900 and 1,500 m (2950–4925 ft) and with local relief often exceeding 300 m (985 ft). The terrain has been an obstacle to transportation in the mountains, historically limiting access; consequently, the western side of the study area has been rather isolated and slow to develop economically (Figure 4.1).

Home to more than 800,000 people (North Carolina Office of State Planning 1999), the twelve-county region has undergone a series of significant transitions over the past three hundred years. First was the transition from a nearly fully forested area of low population density to one of intensive rural agriculture in the early nineteenth century (Silver 1990). Next was the development of an industrial economy during the early twentieth century based largely on textiles and later, on furniture manufacturing (Hayes 1976; Felix and Witcher 1997). In the last decade of the twentieth century, the service sector became a

Figure 4.1 Pilot Mountain in Northwestern North Carolina study area.

major component of the regional economy, with banking and retail sales dominating the Piedmont economy and tourism shaping the Blue Ridge (Finger 1997). Agriculture was a major component of human enterprise in the region throughout its modern history thanks to the high cash value of tobacco grown on the Piedmont and of Christmas trees in the Blue Ridge. Each transition has brought sweeping changes to Northwestern North Carolina's landscapes, and greenhouse gas emissions have changed accordingly.

The Northwestern North Carolina landscape continues to be largely rural, with small land holdings and intermingled fields and woodlots. The study area is dotted with small, scattered towns and cities. Winston–Salem is the only urban area with a population larger than 20,000 within the study area, with a 1990 population of 150,958. Smaller cities are Statesville (17,567), and Lenoir (14,192), Boone (12,949), Kernersville (10,899), Mooresville (9,317), Mount Airy (7,156), Lewisville (6,514), and Clemmons (6,020). In addition, 36 other incorporated places contain fewer than 5,000 people each (United States Department of Commerce, Economic and Statistical Administration, Bureau of Economic Analysis 1997; United States Department of Commerce, Bureau of the Census 1993).

Population in the twelve-county study area increased by 27% from 1970 (582,764) to 1990 (742,484), compared to a 30% increase for the state of North Carolina. The population of the study area increased by 11% between 1990 and 1998, compared to 13.8% for the state (North Carolina Office of State Planning 1999). Most growth occurred in counties with small cities (Watauga and Stokes) rather than in the largest urban county or the mostly rural counties. This growth is largely attributable to the increasing suburbanization of Piedmont counties.

The federal farm-to-market road-building programs of the 1940s and 1950s resulted in a dense net of two-lane highways that crisscross the Piedmont region, linking the rural countryside to nearby towns and to the largest city, Winston–Salem. Interstates 40 and 77 traverse the study area and Interstate 74 is currently being constructed across the region, primarily by upgrading existing divided highways. Northwestern North Carolina's population remains largely rural or suburban, but also largely employed in urban–industrial jobs. These characteristics have created a pattern of modern residential dwellings with large manicured lawns scattered along rural roads throughout the countryside. Many of the rural residents occupying these dwellings continue to reside in the rural communities of their birth, while remaining within what is locally considered to be commuting distance of an urban–industrial job.

Twenty-five miles or more is no longer considered an excessive journey to work in a region where winter driving does not hold the same terrors as in more northerly climes. Long distance commuting to work in a factory has . . . (long been) a fact of life in the South (Hart 1976; Hart and Morgan, 1995).

Suburban and exurban sprawl have greatly affected the Interstate 77 corridor within the study area; large numbers of workers now choose to live in Iredell County and commute to Charlotte. Significant numbers of commuters also travel to work in Winston–Salem from the surrounding counties.

Manufacturing and non-manufacturing enterprises provided approximately equal numbers of jobs in 1970, but manufacturing has declined significantly since then. By 1990, only 32% of all jobs were in manufacturing. The general trend for the state and in the Northwestern North Carolina study area has been a transition away 'from labor-intensive to capital-intensive industry; from manufacturing jobs to trade, service, finance, transportation, and government jobs' (McLaughlin 1997).

Small manufacturing plants are scattered throughout the study area, many of which are subsidiaries or branch plants of larger industries located outside the region (Wheeler 1992). These dispersed locations are attractive to the managers of small plants in such industries as textiles and electronics, who seek inexpensive labor with a strong work ethic, sites that can be developed easily, and, on the Piedmont, a good highway transportation network (Holladay 1997). Many of the plants are also homegrown, particularly in the furniture, food processing, and tobacco industries (McLaughlin 1997).

Growth of these industries through the 1960s was rapid, with ten of the twelve counties having more than 40% of their employment in manufacturing by 1970 (Figure 4.2). These industries provided local employment to a largely rural population, making manufacturing the fastest-growing segment of the economy during that decade. Although manufacturing employment declined relative to other economic sectors between 1970 and 1990, that was largely because other segments of the economy grew more rapidly. More than 30% of total employment remained in manufacturing in ten of the twelve counties in 1990 (United States Department of Commerce, Bureau of the Census 1993).

Several large industries also attracted rural workers seeking industrial jobs after the Second World War. Reynolds Tobacco Company and Hanes Hosiery in Winston–Salem drew rural job-seekers from the surrounding counties. In the latter half of the 1990s, several

Figure 4.2 Furniture manufacturing plant in the Northwestern North Carolina study area.

of the largest corporations in Winston–Salem moved their headquarters elsewhere, owing to a combination of corporate mergers and a desire to site their corporate headquarters in larger cities.

Numerous jobs were lost to developing countries during the late 1990s, a trend that accelerated in the early years of the twenty-first century as the textile industry underwent rapid restructuring. Over a quarter of the textile jobs which existed in North Carolina in 1990 were gone by the middle of 1999 (Hopkins 1999a). Low-wage jobs in the oldest textile mills are currently most vulnerable to additional plant closures, as they are replaced by newer, more efficient plants in Mexico and Asia staffed by lower-wage workers (Hopkins 1999b).

Small farms are scattered throughout the region, but most are part-time operations whose owners work full-time in industrial, commercial, and government jobs in area towns and cities. Consequently, full-time farm employment has declined steadily throughout the past four decades. Tobacco, beef cattle, and poultry are the most important items produced on Piedmont farms, while Christmas trees and tobacco are the largest sources of income from farming in the Blue Ridge.

Greenhouse gas emissions: fossil fuels and forest resources

This diverse landscape of travel corridors, manufacturing plants, and households generates a bundle of the standard greenhouse gases: carbon dioxide, methane, nitrous oxide and ozone-depleting substances. When these gases are assessed on the basis of their hundred-year global warming potential, carbon dioxide is the most important by far. Carbon dioxide was responsible for 89% of the total emissions for the Northwestern North Carolina study area in 1990, compared to 87% for the nation as a whole (United States Environmental Protection Agency 1995a). Methane emissions accounted for 7% of Northwestern North Carolina emissions in 1990, compared to 10% for the United States. Methane emissions are comparatively low because no coal mining, natural gas production, or gas processing takes place within the Northwestern North Carolina study area. In addition, agriculture constitutes a small portion of the economy in the study area, so only minor methane emissions from manure management occur.

Greenhouse gas emissions from the study area are dominated by those associated with the combustion of fossil fuels (Table 4.1). Approximately 90% of the total carbon dioxide equivalent emissions emitted from the Northwestern North Carolina study area in 1990 originated with fossil fuel combustion, if the standard US Environmental Protection Agency protocols are followed. That amount is almost identical to the value for the United States as a whole, but a distinctive feature of the Northwestern North Carolina site not evident in the Environmental Protection Agency protocols is the substantial amount of biomass combustion that occurs within the study area.

The transportation sector accounted for 27.0% of total emissions, the industrial–manufacturing sector for 12.0%, and the residential sector for 5%. Waste disposal systems produced approximately 3.6% of the total greenhouse warming potential, whereas agriculture produced 3.1%, production processes 2.4%, and a variety of other activities generated the remaining 1.5%. Agricultural emissions in the study area were proportionally somewhat lower than the national total, as were emissions of methane from waste disposal (Easterling *et al.* 1998).

Electricity generation is another major source of Northwestern North Carolina greenhouse gases. Utilities generating electricity accounted for 43% of the emissions from the study area in 1990, compared to 36% for the United States. The study area has one large (2,200 MW) coal-fired power plant within its boundaries. Had the study area been designated slightly differently, that plant would have fallen outside its boundaries and the inventory would have had far lower emissions totals. In a small study area, production-based estimates of emissions from utilities may be misleading, depending upon the balance between energy production and consumption within the study area. Electricity generation within the study area occurs at numerous hydroelectric plants, two co-generation plants and the large coal-fired thermal plant (Belews Creek) located in Stokes County and owned by Duke Power Company (Figure 4.3). The amount of electricity used in the study area is almost equal to that produced by the Belews Creek plant. That balance is somewhat misleading, however, in that electricity used within the study area comes from the regional net. More than one third of the regional total of electricity is generated at nuclear and hydroelectric plants, neither of which produce greenhouse gases.

Individual point source emissions of greenhouse gases constitute a dominant proportion of the bundle of emissions from the Northwestern North Carolina study area. The top twenty emitters generated 81% of all Northwestern North Carolina emissions in 1990, with the top ten accounting for 77% of those emissions (Table 4.2). The largest single point source in

Table 4.1 Northwestern North Carolina actual emissions and global warming potential, 1990

Figures are metric tons per year.

Emissions Source Category	CH_4	CH_4 as CO_2 Equivalent	CO_2	N_2O	N_2O as CO_2 Equivalent	ODCs as CO_2 Equivalent	Total CO_2 Equivalent	Percentage of CO_2 Equivalent	Greenhouse Warming Potential
Fossil Fuel Consumption	**1,051**	**22,081**	**16,172,595**	**484**	**130,559**	**0**	**16,325,235**	**89.85**	**16,325,255**
Commercial/Institutional	0	0	523,594	0	124	0	523,718	2.88	523,718
Industrial/Manufacturing	3	70	2,169,645	44	11,868	0	2,181,583	12.01	2,181,583
Residential	5	107	828,712	38	10,183	0	839,002	4.62	839,002
Utilities	49	1,024	7,880,322	90	24,194	0	7,905,540	43.51	7,905,540
Transportation	994	20,880	4,770,322	312	84,190	0	4,875,392	26.83	4,875,392
Biomass Fuel Consumption	**4,388**	**92,143**	**0**	**113**	**30,483**	**0**	**122,626**	**0.67**	**122,626**
Commercial/Institutional	0	0	0	0	0	0	0	0.00	0
Industrial/Manufacturing	24	509	0	22	5,889	0	6,398	0.04	6,398
Residential	4,364	91,364	0	91	24,594	0	116,228	0.64	116,228
Utilities	0	0	0	0	0	0	0	0.00	0
Production Processes	**0**	**0**	**0**	**0**	**0**	**432,281**	**432,281**	**2.38**	**432,281**
Lime Processing	0	0	0	0	0	0	0	0.00	0
Ozone-Depleting Compounds	0	0	0	0	0	432,281	432,281	2.38	432,281
Other Processes	0	0	0	0	0	0	0	0.00	0
Product End Use	0	0	0	0	0	0	0	0.00	0

Agriculture and Livestock

Production	**20,055**	**421,153**	**36,144**	**308**	**83,122**	**0**	**540,419**	**2.97**	**540,419**
Domestic Animals	9,647	202,586	0	0	0	0	202,586	1.12	202,586
Animal Manure Management	10,408	218,567	0	0	0	0	218,567	1.20	218,567
Game Animals (Deer)	0	0	0	0	0	0	0	0.00	0
Fertilizer Use/ Agricultural Liming	0	0	36,144	308	83,122	0	119,266	0.66	119,266
Waste Disposal, Treatment, Recovery	**29,109**	**611,278**	**13,789**	**4**	**1,131**	**0**	**626,198**	**3.45**	**626,198**
Landfills	28,407	596,540	0	0	0	0	596,540	3.28	596,540
Waste Incineration	3	53	13,789	0	0	0	13,842	0.08	13,842
Burning of Agricultural Waste	93	1,952	0	4	1,131	0	3,083	0.02	3,083
Sewage Treatment	606	12,733	0	0	0	0	12,733	0.07	12,733
Human Emissions	**5,509**	**115,689**	**6,493**	**0**	**0**	**0**	**122,181**	**0.67**	**122,181**
Land Use Changes	**0**	**0**	**0**	**0**	**0**	**0**	**0**	**0.00**	**0**
Total Emissions	**60,112**	**1,262,344**	**16,229,021**	**909**	**245,295**	**432,281**	**18,168,940**	**100.00**	**18,168,940**
Percent Global Warming Potential	6.95	6.95	89.32		1.35	2.38	100.00	100.00	100.00

Figure 4.3 Belews Creek coal-fired power plant in the Northwestern North Carolina study area.

the Northwestern North Carolina region is the Belews Creek coal-fired power plant. That facility released one third of all Northwestern North Carolina emissions in 1990. While those production-based emissions occur on behalf of the ultimate end users, it is clear that substantial opportunities to reduce emissions exist at the major production sites.

Industrial biomass combustion

Standard accounting procedures do not include carbon dioxide from biomass combustion, assuming that it is replaced by forest regrowth, which stores carbon (Kates *et al.* 1998). Over time, it should neither add to nor deduct from the bundle of greenhouse gases. If net deforestation is occurring to supply the biomass, such changes in stock should be captured in the land-use component of the inventory (United States Environmental Protection Agency 1995b). International accounting standards, specifically those of the Intergovernmental Panel on Climate Change, do not consider biomass combustion and land-use change to be independent of one another (Intergovernmental Panel on Climate Change 1995). The carbon in trees raised for biomass combustion is inventoried under the category of land use. Carbon being taken up by these trees must be subtracted from carbon in the emissions of biomass combustion in order to correctly compute the net amount of carbon being added to the atmosphere by biomass combustion. Provided that biomass replacement (stand establishment and growth) proceeds at the same rate as biomass harvesting and combustion, then there is no net emission of carbon to the atmosphere by the combustion.

Table 4.2 Major greenhouse gas emitters in the Northwestern North Carolina study area, 1990

Organization	Percentage of Study Area Emissions	Cumulative Percentage of Study Area Emissions
Duke Energy	40.93	41.00
Corn Products	16.60	57.54
R J Reynolds	7.35	64.89
Abtco	4.86	69.75
Appalachian State University	2.90	72.65
Burlington Mills	0.99	73.64
Tyson Foods	0.98	74.62
Tyson Foods 2	0.88	75.50
Kincaid Furniture	0.79	76.29
Chatham Manufacturing	0.74	77.03
Transcontinental Gas	0.64	77.67
Cross Creek	0.64	78.30
Broyhill Furniture	0.54	78.85
Bernhardt Furniture	0.50	79.34
Armtex	0.49	79.83
American Drew	0.43	80.27
Singer Furniture	0.31	80.58
Thomasville Furniture	0.22	80.80
Thomson Crown	0.15	80.96
Nu Woods	0.15	81.10
All Others	18.90	100.00

The Intergovernmental Panel on Climate Change biomass – land use connection is questionable in the Northwestern North Carolina study site primarily because of the local significance of biomass combustion. Including biomass combustion emissions in study area accounts is reasonable in Northwestern North Carolina because substantial portions of the biomass combusted there originate outside the region (see Box 4.1). When the greenhouse gas inventory for the Northwestern North Carolina study area is recomputed with the inclusion of carbon dioxide from biomass combustion, total emissions increase by 20% (see Table 4.1).

One could also argue, however, that from a national perspective, local biomass combustion actually represents a substantial reduction in net emissions because opportunistic plant managers are using a renewable source of energy that reduces fossil fuel consumption. In the case of the largest single user of biomass in the study area, industrial boilers are fired with local wood, right-of-way clearing materials, and furniture waste. The wood chips obtained from those sources have replaced coal (Stanley 1998). As a result, this opportunistic combustion of wood residue results in significant reductions in net carbon emissions from

Box 4.1 Biomass residue fuel to conserve fossil fuels and lower greenhouse gas emissions
Neal G. Lineback

The use of biomass residue as an industrial fuel is a growing business in the southeastern United States. The process of using wood residue in industrial boilers can provide a substitute for fossil fuel energy, conserving non-renewable fuel resources and keeping new carbon from being added to the carbon cycle.

Wood residue from North Carolina's lumber products and furniture industries plus greenwood chips from site clearing and rights-of-way maintenance provide an abundance of potential boiler feed for the state's burgeoning energy needs. Largely driven by biomass residue's availability and relative low cost, as well as new boiler technology, an increasing array of Northwestern North Carolina industries are using this fuel.

Although North Carolina's furniture industry has long combusted its sawdust for space heating and in kilns for drying its wood, other industries are now purchasing surplus amounts of this high-quality wood fuel from furniture manufacturers for use in large boilers. A food processor, a textile mill, and even a hospital now contract for sawdust and other forms of biomass residue.

Other forms include woody landfill residues, wood scrap from building construction and demolition, and greenwood chips from construction sites and highway and power line rights-of-way. Scrap wood must be ground into smaller pieces to be fed into boilers, a process that often takes place at landfill sites. This allows municipalities to charge a tipping fee to handle the materials, as well as a cost per ton for the ground residues.

Greenwood chips are the most abundant wood fuel residue available, but present several problems that make them less valuable than dry wood. For each 1% of moisture in the residue, the energy per unit mass declines by 1%. Wood chips harvested during the summer contain abundant green leaves and wood moisture as high as 40%. Formerly viewed as wood waste, these wood residues increasingly have value as boiler feed. Although combustion of wood residues cycles carbon faster than normal decay, the amount of atmospheric carbon released per unit of mass is approximately the same as through normal decay processes. Thus, no 'new' carbon is entering the active carbon cycle when biomass residue is combusted.

This means that the carbon of an equivalent amount of fossil fuel energy remains sequestered and conserved. The energy ratio between biomass residue and coal, the standard comparative substitute for biomass, is 1 : 0.6 or 1 : 0.7, depending upon the quality of both. With biomass residue selling in 1999 at US$9.00–12.00 per ton and coal at US$35.00–45.00, the cost and carbon advantages of combusting biomass residues are evident.

An issue that has not surfaced to date in North Carolina has been the harvesting of low-value standing timber for industrial energy. Although this practice is occurring elsewhere – most notably in New England – there are environmental concerns about altered species mix and changed wildlife habitat resulting from this practice.

Given the environmental and economic advantages of industrial combustion of biomass residues as a substitute for fossil fuel, this option should be recognized as a mitigation strategy for lowering greenhouse gas emissions.

two forms of avoided activity: (1) the combustion of coal, natural gas, and fuel oil; and (2) perhaps more important, burial of wood residue in landfills. When wood residues are placed into landfills, much of the carbon is converted into methane. The global warming potential of methane is 21, so much higher net radiative forcing would result from the burial of wood residues than from their combustion for local energy (Intergovernmental Panel on Climate Change 1995; Lineback *et al.* 1999).

In sum, the large carbon dioxide emissions from industrial biomass combustion can be accounted for in at least three different ways: (1) as a renewable resource that over time does not contribute to the bundle of greenhouse gas emissions, as it fosters in regrowth as much carbon as it releases through combustion; (2) as an imported source of fossil fuel similar to coal, oil, or gas combusted within the study area and thus a major contributor of greenhouse gases; or (3) as a means of disposal of waste wood that would otherwise release methane and that can substitute for fossil fuel imports, thereby reducing the overall sources of emissions.

Storing carbon in forests

It has been known for some time that mid-latitude forests are sequestering significant amounts of carbon through reforestation (Meyer and Turner 1994). United States Forest Service inventories have shown that biomass is accumulating in North Carolina forests at rates greater than can be explained by reforestation alone (Brown 1993). Such biomass accumulation has been explained as a result of intensive forest management, implying that accelerated carbon accumulation is the result of intentional actions by private woodlot owners (Moffat 1997; Brown 1993). More recently, evidence has emerged which indicates that far faster carbon sequestration is occurring in eastern North American forests than was previously believed. Some researchers have attributed such rapid growth to a combination of carbon dioxide and nitrate fertilization (Fan *et al.* 1998).

The standard Intergovernmental Panel on Climate Change/Environmental Protection Agency protocol for calculating the results of land-use change on emissions requires that forest statistics for only two temporal data points be measured. The amount of biomass contained within those forests, pastures, crops, and soils is then estimated for each time interval and annualized. The resulting number is assumed to represent the amount of carbon that is being released through forest clearing or sequestered through afforestation or reforestation (Intergovernmental Panel on Climate Change 1995; United States Environmental Protection Agency 1995b). A problem with this approach is that short-term trends in land use can result in unreasonably large or small estimates of carbon sequestration or release from

Box 4.2 Forest cover and the 'missing sink' in North Carolina.
Michael W. Mayfield

Global carbon budgets have long indicated that a substantial amount of carbon released from human activities fails to show up in the atmosphere. This 'missing sink'
is generally attributed to accelerated carbon uptake by forests.

Forests in the Northwestern North Carolina study area have undergone dramatic
changes in the past century, and the past three decades have been times of especially
rapid forest biomass change. Essentially all of the forest cover in this area was removed
for timber and for agricultural expansion during the nineteenth and early twentieth
centuries, but forest reversion has been the general trend since the 1950s. Soil depletion
and erosion contributed to a decline in agricultural acreage that was caused primarily
by a shift in the region's traditional agricultural economy, first to light industry, and
later to a service-based economy (Seaver and Mayfield 1995).

Short-term changes in the regional economy and crop prices at the national level
can bring about substantial changes in the regional carbon budget. When standard US
Forest Service forest inventory periods coincide with local trends, unreasonably large
or small estimates of carbon sequestration or release can be from land-use change. The
Northwestern North Carolina study area provides an ideal example of this problem, as
forest inventories were completed in 1984 and 1990.

When the standard Intergovernmental Panel on Climate Change – Environmental
Protection Agency accounting method is applied to these data, startling results emerge.
Because the period from 1984 to 1990 was a time of significant forest reversion, large
amounts of carbon were sequestered in emergent forests. Therefore, a large percentage
of the anthropogenic emissions from the study area were offset by land-use changes.
Such rates are not representative of long-term trends, but rather are an artifact of
the rather substantial forest clearing that occurred in the late 1970s in response to
market conditions. Since 1980, large areas have reverted to forest. The rapid rates
of carbon sequestration associated with that reforestation are not sustainable, for two
reasons: (1) marginal lands have already reverted to forest across the study area; and
(2) biomass accumulation is most rapid in forests that are ten to twenty years old. As
these emergent forests enter older age-class cohort groups, net biomass accumulation
will fall as respiration begins to catch up with net primary productivity.

land-use change. The Northwestern North Carolina study area provides an example
(Box 4.2).

Much of the Northwestern North Carolina study area is forested today, but significant
changes in forest cover have occurred over the past twenty-five years. In 1990, the last year
for which forest statistics are available, 59% of the study area was covered by forest of
all types (Brown 1993). Significant differences in forest cover exist within the study area.
The Piedmont counties have much lower percentages of their land covered by forest than
the mountain counties. For example, Davie, Forsyth, and Iredell counties were less than

45% forested in 1990, whereas Caldwell and Wilkes counties were more than 74% forested (Brown 1993).

Forest cover has changed dramatically since 1974, when 62% of the study area was forested (United States Department of Agriculture Forest Service 1975). Substantial deforestation occurred between 1974 and 1984 due to a combination of market forces and government policies (Rothrock 1999; Henderson and Walsh 1995). Net deforestation in the study area approached 80,000 hectares (200,000 acres) during that interval (United States Department of Agriculture Forest Service 1975, 1986). Most of the deforestation between 1974 and 1984 was concentrated in half of the study area counties. The trend toward forest removal was reversed between 1984 and 1990, as net forest change was approximately 40,000 hectares (100,000 acres) of reforestation (United States Department of Agriculture Forest Service 1986; Brown 1993).

The age-class structure of forests in the Blue Ridge is such that sequestration of carbon has already slowed. The current interpretation of articles adopted at Kyoto compares baseline forest conditions in 1990 to a compliance period from 2008 to 2012 (Schlamadinger 1999). Because rapid reforestation occurred immediately before the 1990 baseline period, it is quite possible that the region will be penalized for accomplishing the reforestation a decade too soon. Research is needed to couple forest inventory statistics with satellite imagery in order to obtain better estimates of the amount of carbon being stored in forests (Pampaloni et al. 1998). Such research would provide finer spatial and temporal resolution than is provided by United States Department of Agriculture Forest Service inventories, which are conducted only about once a decade for entire states or for major physiographic subdivisions of states (Birdsey 1992). Global Change and Local Places estimates of biomass changes at the county level have been limited to inventories of above-ground biomass. Great uncertainty exists regarding the amount of carbon that is in storage below the surface, but recent research supports the idea that soil carbon may be a more important component of forest sequestration than above-ground biomass. In order to claim credit for forest sequestration under the Kyoto accords, it will be necessary to provide much more detailed data.

Driving greenhouse gas emissions: regional growth and change

Temporal trends in greenhouse gas emissions from the Northwestern North Carolina study area reflect a complex, and perhaps contradictory, set of changes. Several of these factors are local in scale, and have served to increase greenhouse gas emissions coming from the study area. Substantial increases in total population, in the number of households, and increasing industrialization within the study area drove emissions higher between 1970 and 1990. But this upward trend in emissions was altered by factors primarily occurring outside the study area, including national improvements in energy efficiency standards after the 1970s energy crisis, and fuel-switching opportunities. For instance, while electricity consumption within the study area was driven higher by increases in people and houses, these increases were partly offset by lower emissions from electricity generation and improvements in the efficiency of energy use.

Population increased by 1.35% per year in the study area between 1970 and 1990, a fact that in itself drives greenhouse gas emissions upwards. The number of households increased from 180,047 in 1970 to 291,106 in 1990, an increase of 45%. An increasing share of the labor force, however, has been employed in the commercial and service sectors of the economy, which produce greenhouse gases at lower rates than manufacturing. Nonetheless, reliance upon private vehicles for commuting and shopping, upon electricity for home heating, and upon trucks for transportation of manufactured and commercial goods, has been driven to a considerable degree by the increase in population.

Net emissions from the study area doubled between 1970 and 1980, primarily because the Belews Creek coal-fired power plant went on line in 1974. Emissions increased by only 7% between 1980 and 1990. When emissions are assigned to the locations of their end users rather than the locations of producers, however, significantly different results are obtained. Utilities, for instance, generate electricity (and thus, greenhouse gas emissions) not for themselves, but for their customers. When an end use analysis of Northwestern North Carolina was done, it was determined that 60.9% of the greenhouse warming potential from the study area originated in the commercial and industrial sectors, while the residential sector was responsible for 36.4% and agriculture only 2.7%. Net emissions from the study area could increase substantially as nuclear power plants now operating on the periphery of the study area are decommissioned (Hamme 1998). If those plants are replaced by fossil-fuel-burning facilities, end-user emissions from the Northwestern North Carolina study area could increase by 50% by the year 2020.

Meanwhile, large increases in the number of registered vehicles, homes, and retail businesses were offset by significant increases in efficiency. For example, the fuel efficiency of the statewide vehicle fleet in North Carolina increased from an average 5.41 km l^{-1} (12.74 mpg) in 1980, to 6.72 km l^{-1} (15.81 mpg) in 1990. The passenger car fleet showed an even greater efficiency increase (Morris 1997). Changes in the sources of energy can have substantial impacts on end-user emission patterns. When the McGuire nuclear power plant was put on line in 1982, end-user emissions dropped by 2.2 million tons (Angel *et al.* 1998).

Methods of household heating vary widely across the study area. Electric heating, however, prevails as the home heating choice (40% of the occupied housing stock), having replaced fuel oil (34%) as the leading heating source between 1980 and 1990. Natural gas is not available in the western counties of the study area, except as bottled propane, but it heats 10% of the homes in the study area. By 2005, natural gas pipelines should be completed to most of the remaining counties, making fuel switching a possibility throughout the study area. Some wood is used for home heating throughout the area, increasing from 4.5% of homes in 1970 to 11.0% in 1980 and remaining at that level through 1990 (Appalachian State University 1996).

The cultural landscape pattern of this mostly rural study area hampers energy efficiency, for it depends almost exclusively upon the automobile and truck for transportation. Except for Winston–Salem and individual county public school transportation buses, public transportation in rural areas is nearly non-existent (Appalachian State University 1996). Consequently, if the automobile is considered an American cultural icon elsewhere, it certainly has strong utilitarian and iconic status among the study area's citizens.

Southerners love trucks (Box 4.3). Trucks accounted for more than 37% of all vehicles registered in North Carolina in 1995, up from only 30% in 1990 (United States

Box 4.3 NASCAR's southern thunder: a greenhouse gas connection.
Neal G. Lineback

The country's fastest-growing spectator sport, NASCAR (National Association of Stock Car Racing) racing, has its roots in the rural South and epitomizes the American love affair with automobiles. Spectators at a NASCAR Winston Cup Race cannot help but be awed by the sound, fury, and pageantry of an event. Cranking out more than 600 horsepower, the specially modified stock cars may reach 200 miles per hour on an oval track in front of 100,000 or more spectators. Such events not only entertain, but also promote the automobile as an American cultural icon. Stock car culture is a Southern tradition so strong that most faithful can identify all qualifying NASCAR drivers by name and by sponsor. Raising the level of the automobile to icon status through stock car racing venues has created a culture that promotes fast cars and 'good old boys.'

Although stock car racing began at small dirt tracks in the Carolinas in the mid-1900s, it was disorganized and unregulated. Often drivers were locally known for their fast cars and highway racing talents or for running moonshine. NASCAR was formed in 1948 in order to bring some organization to a growing racing circuit. The first NASCAR race was held at Daytona, Florida, one year later.

In the early history of NASCAR, events were small and 1940s-vintage cars were mostly stock cars just as they came off the assembly line. Modified cars were permitted during the organization's first few years, but spectators did not seem to relate to them. During WWII, few cars rolled off the automotive assembly lines, so most of the public drove pre-war cars. NASCAR organizer Bill France recognized that, as automotive manufacturers retooled and the economy improved, people would be interested in the new generation of American cars and their performance. Stock car races offered competition between makes and models, as well as specific drivers, leading to fan loyalty to both automobile makes and drivers.

The rise of NASCAR to national prominence grew steadily through the 1960s and 1970s, but the 1980s and 1990s witnessed accelerated growth. Massive corporate support for individual race cars and through advertising by tobacco, soft drink, beer, auto parts, hardware, and home supply industries helped vault NASCAR influence.

No longer involving truly stock automobiles, NASCAR rules allow modifications of basic auto makers' designs to meet strict aerodynamic, structural, and performance standards. The process leaves the car's exterior approximating the original design, but almost everything else is modified. The purpose is to ensure competitiveness and driver and fan safety, thus providing exciting but relatively safe races. These cars, however, are still called 'stock cars,' a holdover from NASCAR's origins.

In the first three decades of existence, NASCAR's fans were mostly from the middle to lower social and economic classes, mainly rural, and white. During the past two decades, however, corporate involvement has brought respectability to stock car racing. The 1990s have seen an expansion of the NASCAR fan pool into the middle and upper economic classes, as corporate executives, lawyers, and physicians have found stock car racing events both entertaining and increasingly fashionable.

Intentionally or otherwise, NASCAR promotes a macho, 'pedal to the metal,' hard-driving, brute-power, fast-automobile mentality among its fans. This is the antithesis of the small engines, fuel efficiency, and fuel economy promoted by those concerned with greenhouse gases and global warming. It is not just that NASCAR events themselves are large producers of carbon dioxide – which they are, if fan transportation is included – but that the NASCAR culture actually discourages fuel conservation and fuel efficiency among its fans and the general public.

NASCAR culture, promoted by corporate sponsorship, is a powerful aphrodisiac for American automobile enthusiasts. Its pageantry, heroes, and automobile icons achieve near-cult status among its most devoted fans, who follow the sport religiously. But even among occasional fans, the NASCAR aura and mystique exert subtle influences on many of their attitudes involving greenhouse gas emissions.

NASCAR is no longer a Southern phenomenon. With its largest tracks scattered in California, Delaware, Indiana, Michigan, New Hampshire, and Texas, as well as Alabama, Georgia and North Carolina, its tentacles of influence reach even into America's Indy heartland. Who would have guessed that stock car racing would achieve national status with the ability to influence such a deep and wide cross-section of Americans? Reversing this influence among NASCAR fans will be a difficult task.

Department of Transportation Federal Highway Administration 1999). Trucks not only are the transportation vehicle of choice (Figure 4.4), but are absolutely necessary in the industrial and commercial sectors. Eighteen percent of the total bundle of the study area's greenhouse gas emissions comes from industrial–commercial transportation sources. Industrial–commercial transportation greenhouse gas sources derive largely from the fact that truck transportation is the only option among the widely scattered small industries and small towns.

Fossil fuel use for transportation was a growing source of greenhouse gas emissions between 1970 and 1990 within the study area, but clearly it is not growing as rapidly as the population. The fact that transportation emissions grew more slowly than population is attributable to increased auto and truck fuel efficiencies (Appalachian State University 1996). Vehicle efficiency leveled off in the mid-1990s as consumer choices in vehicle type changed dramatically. As in most of the nation, sport utility vehicles and vans have become popular, resulting in reduced efficiency of the vehicle fleet. Nationwide, total fleet fuel efficiency declined from 1997 to 2000 after a quarter-century of improvement (United States Department of Transportation 2002).

Agriculture is in decline in the study area, and only poultry production has been able to increase output. Dairy farms in particular have declined in both numbers and animals, in response partly to environmental requirements and partly to increased costs and decreased profits. Beef cattle production, most of which is open grazing with supplemental feeding during the winter, produces little greenhouse gas from beef cattle waste.

Figure 4.4 Big Daddy's Restaurant and Oyster Bar in the Northwestern North Carolina study area.

How do these intermediate driving forces of population and household growth, increasing industrialization, power plants going on line, and changes in the efficiency of automobiles relate to the generic I = PAT formulation of such driving forces? The E = PAT version of the formulation (where E stands for greenhouse gas emissions, P is population, A is a measure of affluence, and T a measure of technology) has been employed by the Appalachian State University Global Change and Local Places research team in an attempt to understand the driving forces behind greenhouse gas emissions in Northwestern North Carolina (Ehrlich and Holdren 1971). While numerous studies have used the I = PAT formulation in an attempt to understand patterns of emissions at the national and global scales (Commoner 1991; Dietz and Rosa 1994, 1997), the Global Change and Local Places project examined its utility at the local level.

To understand the social driving forces of greenhouse gas in the Northwestern North Carolina study area, both consumption-based (end-user) and production-based (actual) measures of emissions were statistically compared to socioeconomic characteristics at the county level for 1990 (Soulé and DeHart 1998). Numerous measures of population and affluence were related to total emissions within the study area and to sectoral (residential, commercial–industrial and agricultural) emissions, measured in global warming potential. Consumption-based measures of greenhouse gas emissions were found to be more strongly related to population- and affluence-based variables than production-based measures of greenhouse gas emissions (Soulé and DeHart 1998: 759).

Changes in the driving forces of greenhouse gas emissions in the Northwestern North Carolina study area were examined by comparing production-based measures of greenhouse gas emissions to socioeconomic characteristics for the decades of 1970, 1980, and 1990. Population-based socioeconomic characteristics were strongly related to all emissions categories except for agricultural emissions. A nuclear power facility came on line during the 1980s, dramatically reducing the area's reliance on coal-fired electric generation. This shift in fuel source influenced the statistical relationships between emissions and socioeconomic characteristics for the 1980–1990 period. Over the longer period of 1970–1990 there existed a stronger relationship between emissions and population- and affluence-based characteristics.

A large portion of the increase in greenhouse gas emissions is likely related to increases in the number of vehicles and subsequent miles traveled, increases in the number of households, increases in population, and increasing industrial output (DeHart and Soulé 2000).

Reducing greenhouse gas emissions: technology potentials and the quality of life

To gauge local attitudes toward potentials to reduce greenhouse gas emissions from the Northwestern North Carolina study area, both representatives of major emitters and a sample of households were surveyed. The survey results reveal the ways Northwestern North Carolina officials and residents perceive the existence of global warming resulting from greenhouse gas emissions and how they might change their behavior in response to those perceptions.

Major emitters

Officials representing twenty major greenhouse gas emitters (accounting for 65% of all study area emissions) were interviewed in late 1998 and early 1999 to ascertain their levels of awareness and concern over greenhouse gas issues, determine the locus of control over corporate decision-making that could be used in mitigation efforts, learn of technological or operational opportunities for reducing emissions at the sites, and find out which regulatory agencies are considered to be most effective. A wide range of attitudes regarding the importance of greenhouse gas emissions was found among the interviewees.

The industries emitting the largest quantities of carbon dioxide were most keenly aware of greenhouse gas issues and most concerned about the Kyoto Accord. Duke Energy officials have attended all of the major global warming conferences and lobby actively for regulations that they feel to be effective but reasonable. R. J. Reynolds (RJR) personnel were equally familiar with regulatory and scientific issues but appeared to be more skeptical about the reality of global warming. RJR was quite active in lobbying the United States Senate to reject the Kyoto Protocol. Piedmont Natural Gas Company officials recognize that natural gas suppliers stand to gain significantly from any effort to reduce carbon emissions, but were nonetheless skeptical about the reality of global warming and about government actions to avoid potential impacts of carbon emissions. Several major emitters were members of industry groups that have opposed implementation of the Kyoto Protocol.

Furniture manufacturers and Appalachian State University boiler plant officials were far more knowledgeable about and concerned with air quality issues than with greenhouse gas emissions. Examples include recent modifications to the Clean Air Act (CAA) and a recent Environmental Protection Agency ruling that will require substantial reductions in nitrous oxide emissions.

Several key factors may radically affect the success of future mitigation strategies in the Northwestern North Carolina study area: (1) future decommissioning of both nuclear and coal-fired thermoelectric generating plants; (2) abandonment of antiquated factories and the building of newer, more efficient plants; (3) the increasing size and decreasing efficiency of private vehicles; and (4) trends toward larger private residences and consequent increased heating and air-conditioning costs. Although these factors are not specific to this region, they may have an enormous potential effect on attempts to reduce local area greenhouse gas emissions.

Approximately 30% of the electricity used in the study area derives from seven nuclear thermoelectric units located in North and South Carolina, three of which will reach their projected thirty-year life expectancies by 2012. In addition, of the 27 North Carolina coal-fired thermoelectric units, 20 will have met their fifty-year life expectancies by 2012. Unless the power companies successfully petition to extend their lives, these plants will be decommissioned. In the case of nuclear plant decommissioning, net carbon emissions will increase regardless of the type of generating capacity that replaces the decommissioned facilities. All replacement generating options in the near to mid-term (ten to thirty years) will generate substantial amounts of carbon dioxide.

Decommissioned plants will most likely be replaced with gas-fired turbines in the study area, at least through 2010. Replacing nuclear generating capacity with gas turbines would increase the overall global warming potential emissions of the Northwestern North Carolina study area by at least 7%. Replacing all of the decommissioned plants and meeting increased demand with gas turbines will be difficult because the existing gas pipeline infrastructure is inadequate to transport the necessary gas. However, new gas pipelines are being planned throughout the study area. Replacement of older coal-fired generators with natural gas units will further strain delivery capacity. Opposition to construction of new generating plants by local communities is likely to result in gas-fired units replacing coal-fired units at existing locations.

Spatial adjustments in manufacturing patterns in the Northwestern North Carolina study area involve the abandonment of antiquated factories, substitution of more efficient for outmoded industrial technologies, and relocation of some industries outside the region. For instance, adjustments in the textile industry involve all of the above, as NAFTA has increased economic pressures on North Carolina's aging textile plants.

The tobacco industry is also under pressure to downsize and increase efficiencies, and has begun moving in those directions. Replacement of older motors with high efficiency motors, electricity co-generation, and fuel switching are some of the obvious strategies, but overall savings from these are relatively small in comparison with the industry's other fixed and operating costs.

Other industries also seek greater energy efficiencies. Use of the abundantly available biomass residue as an industrial fuel is a trend whose future depends greatly on acceptance by the United States Environmental Protection Agency of biomass as a recognized substitute for burning fossil fuels. Switching from oil and coal to natural gas by industry is a recognized strategy, but again the existing infrastructure is not adequate to supply the volume of natural gas required.

Household perspectives

A survey of local attitudes regarding greenhouse gas emissions and mitigation potentials in the study area indicated a wide range of knowledge about greenhouse gas issues and a wide range of willingness to mitigate. An opinion survey of 843 households in the twelve-county Northwestern North Carolina study area dealt with knowledge base, attitudes, responsibilities, and interest in mitigation strategies. The households were chosen randomly and the return rate averaged approximately 28% across the entire study area. These respondents were primarily older and wealthier than the median adult residents: 81.4% were 35 years of age or older and 64.4% had annual incomes greater than US$30,000.

Most respondents believed that fossil fuel use caused global warming and assumed that individual households and the federal government should be responsible for any reductions. However, most believed that lack of knowledge about global climate change was responsible for not implementing measures to reduce greenhouse gas emissions.

Attitudes of the respondents generally reflect the prevailing rural lifestyle and the scattered industrial sites across the study area. Many Northwestern North Carolinians have been able to maintain their neighborhood and family ties by living in their rural communities of birth while working in industrial and commercial jobs in nearby towns and county seats. The factors that have made this possible have been growth in the number of local jobs that has kept close pace with population increase, a good secondary road system crisscrossing the region, and reliance on passenger vehicles. Thus, the respondent households averaged 2.26 vehicles each, with nearly half owning a truck or sport utility vehicle. Seventy percent of the respondents believed their vehicles to be fuel-efficient, and 53% would purchase a more fuel-efficient vehicle. Weekly driving distances per household averaged 180 miles across the study area, attributed to lengthy journeys to work, the rural nature of the region, and the lack of public transportation.

Across the study area, home heating and cooling are major issues. Forty-two percent of the respondents heated their homes with electricity and 21% by fuel oil. Because natural gas is not available in much of the study area, fuel-switching potentials are somewhat limited. Respondents expressed a strong willingness to increase energy efficiency by installing weather-stripping and storm doors in their homes, as well as hanging window shades, using window tinting, and planting trees.

Prospects

Gross carbon dioxide equivalent emissions are expected to grow significantly in Northwestern North Carolina by 2010, based upon business-as-usual projections (Appalachian

State University 2000). Mitigation of greenhouse gas emissions in the Northwestern North Carolina study area may require a 40% reduction by 2010 to meet the 7% below 1990 standard proposed at Kyoto. Assuming this business-as-usual scenario, the Northwestern North Carolina study area should generate 24.2 million tons of carbon dioxide equivalent in 2010, up from 18.1 million in 1990. Over the longer term to 2020, a 20% below 1990 reduction standard would require a 50% or greater reduction in carbon dioxide equivalent. How might such large cuts in greenhouse gases be accomplished?

Population growth, vehicle choice, a rapid increase in vehicle miles traveled, and housing characteristics have led to a significant increase in net emissions from the Northwestern North Carolina study area in the late 1990s. Meeting the goals of the Kyoto accords will be extremely difficult. Emissions will, in fact, continue to increase significantly through 2010 in the absence of financial incentives or regulatory requirements to reduce emissions. Utilities and transportation offer the greatest opportunities for emissions reductions, although industrial and domestic end users must also be included in any attempt to reduce emissions. With carbon emissions permits in place, fuel switching at power plants from coal to natural gas would become much more appealing to utilities (Romm *et al.* 1998). Such financial incentives would also accelerate expansion of the natural gas pipeline network and associated infrastructure, providing opportunities for emissions savings from a wide variety of operations.

Carbon permits costing US$50.00 per ton would increase the cost of gasoline by approximately US$0.03 per liter or US$0.125 per gallon (Romm *et al.* 1998). Recent price increases of more than US$0.048 per liter (US$0.20 per gallon) in six months have had no discernible impact on driving behavior in the region, so there is little evidence that such a tax would significantly curtail driving. The rural and suburban character of Northwestern North Carolina encourages large numbers of vehicle trips and long driving distances. New urbanism and other trends toward clustered semi-urban housing could reduce vehicular emissions growth only if coupled with improvements in public transit.

The old industrial economy of the North Carolina Piedmont is in decline, as textile and furniture jobs migrate to low-wage offshore locations. New jobs in the region are being generated largely in low-emissions service activities. That trend will most likely continue in the textile industry and accelerate in the furniture industry, resulting in reduced industrial emissions from the region.

Transformations in the economy of Northwestern North Carolina have brought about dramatic changes in the way that the people of the region live and work. Changes in the economy and the landscape of the region over the next twenty years could be as dramatic as the change from a farm economy to an industrial economy in the early twentieth century.

REFERENCES

Angel, D. A., S. Attoh, D. Kromm, J. DeHart, R. Slocum, and S. White. 1998. The Drivers of Greenhouse Gas Emissions: What Do We Learn from Local Case Studies? *Local Environment*, **3** (3): 263–78.

Appalachian State University. 1996. *The North Carolina Greenhouse Gas Emissions Inventory for 1990*. Boone, NC: Appalachian State University Department of Geography and Planning.

Appalachian State University. 2000. *North Carolina's Sensible Greenhouse Gas Reduction Strategies*. Boone, NC: Appalachian State University Department of Geography and Planning. http://www.geo.appstate.edu/projects/ ncaction/into.html

Birdsey, R. A. 1992. *Carbon Storage and Accumulation in United States Ecosystems*. Radnor, PA: United States Forest Service. USDA Forest Service General Technical Report WO 59.

Brown, M. J. 1993. *North Carolina's Forests, 1990*. USDA Forest Service Technical Report. Asheville, NC: United States Forest Service.

Commoner, B. 1991. Rapid Population Growth and Environmental Stress. *International Journal of Health Services*, **21** (2): 199–227.

DeHart J. L., and P. T. Soulé. 2000. Does I = PAT Work in Local Places? *The Professional Geographer*, **52**: 1–10.

Dietz, T., and E. A. Rosa. 1994. Rethinking the Environmental Impacts of Population, Affluence and Technology. *Human Ecology Review*, **1** (2): 277–300.

Dietz, T., and E. A. Rosa. 1997. Effects of Population and Affluence on CO_2 Emissions. *Proceedings of the National Academy of Sciences*, **94**: 175–9.

Easterling, W. E., C. Polsky, D. Goodin, M. Mayfield, W. A. Muraco, and B. Yarnal. 1998. Changing Places, Changing Emissions: the Cross-Scale Reliability of Greenhouse Gas Emission Inventories in the US. *Local Environment*, **3**: 247–62.

Ehrlich, P. R., and J. P. Holdren. 1971. Impact of Population Growth. *Science*, **171**: 1212–17.

Fan, S., M. Gloor, J. Mahlman, S. Pacala, J. Sarmiento, T. Taahashi, and P. Tans. 1998. A Large Terrestrial Sink in North America Implied by Atmospheric and Oceanic Carbon Dioxide Data and Models. *Science*, **282**: 442–6.

Felix, L., and B. Witcher. 1997. Greenhouse Gas Emissions from the Furniture Industry in Northwest North Carolina. *Papers and Proceedings of the Applied Geography Conference*: 35–42. Albuquerque: University of New Mexico.

Finger, B. 1997. Making the Transition to a Mixed Economy. *North Carolina Insight*, **17**: 2–3, December.

Hamme, R. E. 1998. Manager, Environment, Health, and Safety Issues, Duke Energy. Personal interview, October 2.

Hart, J. F. 1976. *The South*. New York: D. Van Nostrand Company.

Hart, J. F. and J. T. Morgan. 1995. Spersopolis. *The Southeastern Geographer*, **25**: 2.

Hayes, C. R. 1976. *The Dispersed City*. Chicago: University of Chicago Department of Geography Research Paper 173.

Henderson, B. M. and S. J. Walsh. 1995. Plowed, Paved, or in Succession: Land Cover Change on the North Carolina Piedmont. *Southeastern Geographer*, **35** (2): 132–49.

Holladay, J. M. 1997. Five Trends that Strengthen Economies. *North Carolina Insight*, **17**: 2–3.

Hopkins, S. M. 1999a. Nimble California Mills Reshape Textile Trade. *The Charlotte Observer*, May 23: 1A, 12A.

Hopkins, S. M. 1999b. 2,420 Carolinas Textile Jobs Cut. *The Charlotte Observer*, January 27: 1A, 11A.

Intergovernmental Panel on Climate Change, Organization for Economic Cooperation and Development. 1995. *Greenhouse Gas Inventory Workbook*. London: United Nations Environment Program, The Organization for Economic Co-operation and Development, and the Intergovernmental Panel on Climate Change.

Kates, R. W., M. W. Mayfield, R. D. Torrie, and B. Witcher.1998. Methods for Estimating Greenhouse Gas Emissions from Local Places. *Local Environment*, **3** (3): 279–97.

Lineback, N. G., T. Dellinger, L. F. Shienvold, B. Witcher, A. Reynolds, and L. E. Brown. 1999. Industrial Greenhouse Gas Emissions: Does CO_2 From Combusting of Biomass Residue Really Matter? *Climate Research*, **13**: 221–9.

McLaughlin, M. 1997. Trends in the North Carolina Economy: An Introduction. *North Carolina Insight*, **17**: 2–3.

Meyer, W. B. and B. L. Turner II, eds. 1994. *Changes in Land Use and Land Cover: A Global Perspective*. Cambridge: Cambridge University Press.

Moffat, A. S. 1997. Resurgent Forests Can be Greenhouse Gas Sponges. *Science*, **277**: 315–17.

Morris, M. 1997. United States Department of Energy, Energy Information Agency. Personal communication.

North Carolina Office of State Planning. 1999. http://www.ospl.state.nc/

Pampaloni, P., S. Paloscia, and G. Macelloni. 1998. The Potential of C- and L-Band SAR in Assessing Vegetation Biomass: The ERS-1 and JERS-1 Experiments. *Proceedings of the 1998 European Remote Sensing Symposium*. Paris: European Space Agency.

Romm, J., M. Levine, M. Brown, and E. Peterson. 1998. A Road Map for U. S. Carbon Reductions. *Science*, **279**: 669–70.

Rothrock, K. 1999. An Analysis of Land Cover Change in Wilkes County, North Carolina, 1972–1990. Unpublished master's thesis. Boone, NC: Appalachian State University.

Schlamadinger, B. 1999. Status of Chapter 4, IPCC Special Report on Land-Use Change and Forestry, *Proceedings of the IEA Bioenergy Conference*. Graz: Joanneum Research.

Seaver, H. C. and M. W. Mayfield. 1995. Historical Land Use and Accelerated Soil Erosion in Watauga County (1950–1988). *The North Carolina Geographer*, **4**: 31–9.

Silver, T. 1990. *A New Face on the Countryside: Indians, Colonists, and Slaves in South Atlantic Forests, 1500–1800*. Cambridge: Cambridge University Press.

Soulé, P. T., and J. L. DeHart. 1998. Assessing IPAT Using Production- and Consumption-Based Measures of I. *Social Science Quarterly*, **79** (4): 54–765.

Stanley, J. 1998. Plant Engineer, Corn Products, Inc. Personal interview, October 9, Winston–Salem, N. C.

United States Department of Agriculture Forest Service. 1975. *Forest Statistics for the Mountain Region of North Carolina, 1974*. Washington: US Government Printing Office. USFS Resource Bulletin SE-31.

United States Department of Agriculture Forest Service. 1986. *Forest Statistics for North Carolina, 1984*. Washington: US Government Printing Office. USFS Resource Bulletin SE-78.

United States Department of Commerce, Bureau of the Census. 1993. *Census of Population, 1990*. Washington: Government Printing Office.

United States Department of Commerce, Economic and Statistical Administration, Bureau of Economic Analysis. 1997. *Regional Economic Information System (REIS) 1969–95* (CD-ROM). Washington: Government Printing Office.

United States Department of Energy. 2002. http://www.eia.doe/emeu/mer/txt/mer1-10

United States Department of Transportation, Federal Highway Administration. 1999. http://www.bts.gov/ntda/fhwa/prod.html

United States Environmental Protection Agency. 1995a. *Inventory of U. S. Greenhouse Gas Emissions and Sinks: 1990–1994*. Washington DC: U. S. Environmental Protection Agency, Office of Policy, Planning, and Evaluation. EPA-230-R-76-006.

United States Environmental Protection Agency. 1995b. *State Workbook; Methodologies for Estimating Greenhouse Gas Emissions,* 2nd edition. Washington: U. S. Environmental Protection Agency, Office of Policy, Planning, and Evaluation, State and Local Outreach Program. EPA-230-B-92-002.

Wheeler, J. O. 1992. The Changing South: Changing Corporate Control Points in the U. S. South, 1960–1990. *The Southeastern Geographer,* **32** (1), May 1992.

5

Northwestern Ohio: re-industrialization and emission reduction

Samuel A. Aryeetey-Attoh, Peter S. Lindquist, William A. Muraco, and Neil Reid

Toledo and its region attracted the attention of the Global Change and Local Places project because of their centrality in the traditional urban–industrial heartland of the United States. The area has undergone profound changes during the Global Change and Local Places study period. Its heavy involvement in fossil-fuel use in industry and the greenhouse gas emissions from such establishments place Northwestern Ohio at the core of the national search for ways to reduce greenhouse gases.

From 1970 to 1990, the 23-county study region of Northwestern Ohio accomplished a 26% decline in greenhouse gas emissions because of a complex combination of technological and structural forces. Perhaps the key ingredient was the industrial sector's responses to federal and state enforcement of more stringent air pollution controls mandated by the Clean Air Act. A second factor resulting in reduced emissions was local industrial restructuring brought about by changes in the broad region's general economy. Intensive intra-regional competition developed in the 1990s between *greenfield* industrial developments in the rural and suburban counties and to a lesser degree renewal efforts for *brownfield* redevelopment in core urban areas (Bielen 1998). That regional competition led to redistribution of greenhouse gas emissions from core urban zones to suburban and rural areas. Industrial restructuring will most likely continue to determine the magnitudes and locations of future greenhouse gas emissions in Northwestern Ohio. Adoption of energy-efficient production technologies will temper growth in greenhouse gas emissions despite continued industrial growth.

Landscape, life, and livelihood

Northwestern Ohio's economy is rooted in a mixture of urban and rural livelihoods, although the locality's image and economy are dominated by metropolitan Toledo's industrial complex, which in turn relies on the area's location with respect to national and international markets (Figure 2.5). Located at the intersection of Interstates 75 and 80–90, the City of Toledo is situated within 800 km (500 miles) of 38% of the United States population and 35% of Canada's people (Toledo Area Chamber of Commerce 1991). Other large metropolitan areas within 800 km include Chicago, Cincinnati, Cleveland, Columbus, Detroit, Pittsburgh, and Toronto. The region is linked to world markets by the Great Lakes – St. Lawrence

Figure 5.1 BPAmoco refining facility in the Northwestern Ohio study area.

Seaway. The Port of Toledo ships more bulk cargo than most other Great Lakes ports. BAX Global, one of the country's leading air cargo companies, operates from the Toledo Express Airport. Toledo's air and water accessibility is buttressed by its role as one of the country's major rail nodes, served by Conrail, CSX Transportation, and the Norfolk Southern, among others. Positioned just south of Detroit, the Toledo region has a strong manufacturing history, much of it based on automobile and automotive parts assembly. Automotive parts production, industrial machinery and electronic equipment manufacture, oil refining, chemical production, food processing, steel making, and glass production dominate the region's industrial portfolio (Figure 5.1). The region underwent considerable economic restructuring during the 1980s, a period when local dependence on service sector jobs increased. The region's agricultural landscape is dominated by mixed grain farming (especially corn and soybeans) and livestock production (Toledo Area Chamber of Commerce 1991).

The Northwestern Ohio study area had a population of 1,650,100 in 1990, residing in 23 counties covering 13,985 square kilometers (5,400 square miles) in aggregate. The 1990 labor force consisted of 623,894 workers. The majority worked in the service sector, with about 32% employed in manufacturing. By comparison, in the 1970s the majority worked in manufacturing. Toledo dominates the area's urban population with 318,000 people, followed by Mansfield with 174,000 and Lima with 155,000.

Table 5.1 Greenhouse gas emissions for Northwest Ohio and Ohio, 1970–1990

Figures are tons of greenhouse gas.

Area	1970	1980	1990	Percentage Change 1970–1990	Percentage Change 1980–1990
Northwestern Ohio	54,342,000	46,842,000	40,260,000	−25.9	−14.1
Ohio	345,566,000	334,426,000	307,077,000	−11.1	−8.2

Table 5.2 Processes contributing greenhouse gas emissions in Northwestern Ohio, 1990

Process	Percent Contribution
Fossil Fuel Consumption	78.0
Production Processes	8.8
Agriculture	7.2
Waste Disposal, Treatment, and Recovery	4.6
Other	1.4
Total	100.0

Greenhouse gas emissions decline from 1970 to 1990

Northwestern Ohio is distinct from the other three Global Change and Local Places study areas in that greenhouse gas emissions in the area decreased by 25.9% from 1970 to 1990 (Table 5.1). While statewide emissions also dropped by 11.1% during the period, the study area and the decline in its emissions were intriguing to the Global Change and Local Places team because those decreases contrast sharply with the increases observed in the other three Global Change and Local Places study areas.

In 1990, approximately 85% of the region's greenhouse gas emissions were carbon dioxide, 10% were methane, and 2.6% nitrous oxides. By comparison, 90% of Ohio's greenhouse gas emissions were carbon dioxide, 6.4% were methane, and 1.1% were nitrous oxides. Northwestern Ohio more closely approximates the national (86% carbon dioxide, 12% methane, and 2% nitrous oxides) than the Ohio mix. Fossil fuel consumption was responsible for 78% of Northwestern Ohio's greenhouse gas emissions in 1990. Other sources were industrial production, agriculture, and waste disposal, treatment, and recovery (Table 5.2).

In 1990, nearly 40% of study area carbon dioxide came from fossil fuel burning in manufacturing, one third originated in transportation, 16% from utilities, and one tenth from residential heating (Table 5.3). Statewide, utilities emitted 43% of emissions, whereas the manufacturing and transportation sectors accounted for about one fourth and one fifth, respectively. Electricity generated at the Davis Besse nuclear plant located near Sandusky

Table 5.3 Greenhouse gas emissions from
fossil fuel use in Northwestern Ohio and Ohio,
1990

	Percentage of Emissions	
Source	Northwestern Ohio	Ohio
Industry	39.0	23.2
Transportation	32.5	21.0
Utilities	15.7	43.4
Residential	10.3	8.2
Commercial	2.5	4.1
Total	100.0	100.0

meets 44% of Northwestern Ohio's needs, resulting in modest utility greenhouse gas emissions in the study area compared to the state, and to the entire country. Nationally, 89% of
greenhouse gas emissions derive from fossil fuel burning, much of it in coal-fired electricity
generation plants, and only 1% of emissions originate in production processes. The Ohio
study area's access to power from nuclear fuel and its focus on manufacturing and production processes yield an emissions profile distinct from that of the other Global Change and
Local Places study areas.

The dominance of fossil fuels related to manufacturing in the study area's emissions
comes as no surprise given the region's heritage as part of the United States once known as
the Manufacturing Belt. Indeed, steel production continues to release more carbon dioxide
in Northwestern Ohio than any other single source, even though such emissions decreased
from 9,116,000 metric (10,046,000 short) tons in 1970 to 7,783,000 metric (8,577,000
short) tons in 1990, largely as a consequence of the shift from furnaces fueled by coke
to electric furnaces. Automotive assembly and automobile parts production complement
primary metals production in the area, but their greenhouse gas output is modest, standing
at only 350,000 metric (385,000 short) tons of carbon dioxide in 1990.

Between 1970 and 1990, major changes in petroleum refining and production with greenhouse gas implications took place in the study area. Refiners, such as BPAmoco, Sunoco,
and others, released 10,340,000 metric (11,395,000 short) tons of carbon dioxide in 1970.
Large investments in more efficient refining technologies resulted in a 76% reduction in oil
industry emissions in the study area by 1990, to about 7,858,000 metric (8,660,000 short)
tons of carbon dioxide in 1990. After manufacturing and production processes, which together released 39% of the study area's 1990 emissions, the transportation sector makes the
second largest contribution (25%) to Northwestern Ohio's emissions. Just-in-time delivery for auto assembly continues to stimulate the growth of trucking and rail transportation
in the area (Ansari and Modarress 1987). Moreover, changes in the locations of employment will likely result in longer commutes for study area employees (Freeman 1998). The
net effects of both *greenfield* development (relocation of new manufacturing activities in
industrializing rural and suburban counties) and *brownfield* redevelopment (reinvestment

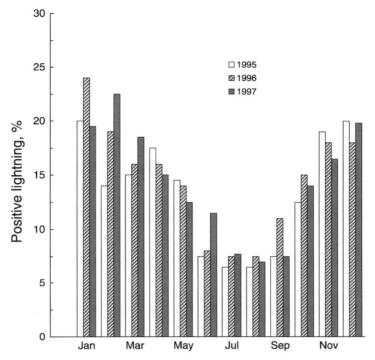

Figure 5.2 Old Jeep plant brownfield site in the Northwestern Ohio study area.

in older urban industrial sites) is greater urban sprawl and longer trips between home and work (Figures 5.2 and 5.3).

Greenhouse gas emissions drivers: technological changes and economic restructuring

Northwestern Ohio was able to reduce greenhouse gas emissions from 1970 to 1990 and especially in the 1980s by investing in more energy-efficient manufacturing and processing technologies, and by shifting to lighter, cleaner manufacturing and to services as part of the major restructuring of the old manufacturing belt region.

Technological innovations directed at emission reductions

The dramatic decline in emissions observed in Northwestern Ohio partly reflects changes in a few key industrial sectors, especially petroleum refining and chemical manufacturing. In these industries and others, reductions in equivalent carbon dioxide emissions due to improved energy-efficient technologies were achieved concurrent with production increases. The stimulus for emission reduction was new environmental standards imposed by federal and state regulations. BPAmoco, the study area's largest oil refiner, invested more than

Figure 5.3 Daimler Chrysler greenfield development in the Northwestern Ohio study area.

US$450,000,000 to install and maintain low-emission, energy-efficient technology. BPAmoco's senior managers are committed to achieving International Standards Organization 14001 Environmental Management System (EMS) certification, and BPAmoco participates in the voluntary Ohio Prevention First Initiative at the Industrial Leadership level, which required a 50% reduction of levels of chemical releases by 2000.

Part of the decline in greenhouse gas emissions was accomplished by technological improvements at the BPAmoco Lima Chemical Plant in Allen County. Carbon dioxide emissions from the plant have been brought down from 7.72 kilograms per kilogram (3.5 pounds per pound) of product in the 1980s to 1.68 kilograms per kilogram (0.76 pounds per pound) of product. Major capital investments have also curtailed emissions in the steel industry. Armco Mansfield, for example, produces secondary steel and alloys from scrap in electric arc furnaces at its plant in the southeast part of the study area. The company's technological investments include the installation of continuous computer casters, which take molten steel directly to slabs, thus eliminating processes that generate carbon dioxide emissions.

National and regional economic restructuring

Northwestern Ohio underwent a major economic restructuring during the 1980s. The region's total labor force grew from 554,000 in 1980 to 624,000 in 1990, an increase of 12.6%.

During the same period, employment in manufacturing in the region declined from 41 to 32% of the labor force. As manufacturing employment dropped, the number of service jobs increased. Service sector employment grew by 25% in the 1980s and the service sector share of the study area labor force rose from 58.1% in 1980 to 66.6% in 1990.

The restructuring was accompanied by a rise in productivity. Though manufacturing employment in Northwestern Ohio decreased by 10.5% from 1980 to 1990, value added by manufacturing rose by 18.2% over the same period. Value added per worker in manufacturing increased from US$17,985 to US$23,760 in the decade, an increase of 32.1%. Global climate change skeptics often allege that greenhouse gas emissions reductions will necessarily result in productivity losses. That has not been the experience in Northwestern Ohio (Box 5.1). In study area counties where greenhouse gas emissions declined, manufacturing productivity increased by 36.9% between 1980 and 1990 compared with the 26.4% growth in study area counties where greenhouse gas emissions increased.

Part of Northwestern Ohio's reduction in greenhouse gas emissions in the 1980s followed changes in the area's industrial base. The most traumatic blow to the area economy was the closing of the Libby – Owens Ford glass plant in 1982, which cost 2,500 jobs. Other manufacturing employers shrunk local operations, Champion Spark Plug company among them. Many of the manufacturing operations that ceased or reduced production and employment in the 1980s were old facilities located in central Toledo. While these older establishments were shrinking or dying, new facilities were replacing them, often located in rural areas and smaller urban centers. Between 1980 and 1990 the number of manufacturing plants in the region actually increased by 10.6%, from 2,492 to 2,756. The new plants are more friendly to the environment than the old establishments, but their scattered locations require longer trips on the part of workers and suppliers, and therefore increased greenhouse gas emissions from the transportation and household (commuting) sectors. Also of note with respect to the locus of control over study area events is the importance of foreign investment in Northwestern Ohio's restructuring (Box 5.2).

Sectoral driving forces

Transportation

Increases in greenhouse gas emissions originating in the transportation sector have partly offset decreased emissions arising in manufacturing, specifically by an upswing in residential vehicle emissions resulting from an increase in miles traveled by Northwestern Ohio residents. During the 1980s, the number of households grew by 4%, with a nearly proportional decrease in the number of persons per household and an even larger increase (14%) in the number of single-person households. Over the same period, the number of women in the workforce increased by 16%, increasing the number of multi-car families. Total registered vehicles in the study area (including passenger cars and non-commercial trucks) rose by 20% during the period and the number of two-vehicle households increased by 10%.

Box 5.1 From rustbelt to greenfields: the story of a region's revival

During the 1960s and 1970s a profound locational realignment of population and industry occurred in the Midwestern and northeastern states of the country that had for decades constituted the geographic core of the national economy. With the national population increasingly concentrated in southeastern and southwestern states, the American Midwest experienced a reversal of its historic prosperity (Gober 1993). The Northwestern Ohio study area was far from immune to the macro-economic and geographic forces at work in the state and throughout the Midwest. The number of manufacturing jobs in the United States increased by 0.3% between 1972 and 1982, but in both Ohio and the study area manufacturing jobs decreased by more than 17%. By the 1980s, however, the Midwest's manufacturing economy had begun to rebound. Manufacturing jobs decreased by 4.7% for the entire country between 1982 and 1992, and Ohio lost 5.8% of its manufacturing employment in that decade. Northwestern Ohio, on the other hand, posted a 1.8% gain in manufacturing jobs for the decade.

Northwestern Ohio's rebound from the rustbelt woes of the 1970s and early 1980s was spurred by an array of business-minded regulatory and training incentives initiated at the state and regional levels. The State of Ohio's Job Creation tax credit program, for instance, offered companies credits against state corporate and income taxes for new jobs. Additional incentives included Manufacturing Machinery and Equipment investment tax credits, Community Reinvestment Area tax exemptions, Enterprise Zone tax exemptions, flexible utility incentive rates, funds for infrastructure improvements, and direct loans and bond financing. These incentives combined with manufacturing and technology education programs to encourage newer, clean, more efficient, and more automated industries to locate in the region.

Inevitably, part of Northwestern Ohio's recovery was linked to the automobile industry. Over the years, automobile manufacturing, assembly, and parts manufacture had experienced growing competition from overseas manufacturers, especially when Japanese manufacturers began to penetrate the United States market. Ironically, the same Japanese automakers that were blamed for the Midwest's hard times in the 1970s are now credited with stimulating the region's revitalization during the 1980s and 1990s (Klier 1981; Ballew and Schnorbus 1987; Bergman and Strauss 1994).

Simultaneous increases in commuting distances reflected Northwestern Ohio's increasing suburbanization. The number of workers commuting to jobs outside their communities of residence grew by almost 30%. Driving habits also changed from 1980 to 1990. The number of commuters in car pools dropped by more than 30%; consequently, the number of drivers traveling alone to work increased by 21%. The proportion of workers using public transit on a regular basis in 1980 had dwindled to 30% by 1990. All of these changes put more household vehicles on Northwestern Ohio's roads, driving longer distances, and

---

Box 5.2 Greenfield development and foreign direct investment

Economic development agents promoting investment in the region aggressively sought foreign plants in the 1980s and 1990s. A huge influx of Japanese direct investment (JDI) into the United States occurred in the 1980s, with Ohio being a favored destination (Kenney and Florida 1991). More than sixty manufacturing facilities were constructed or acquired in Ohio by Japanese, mostly automotive-related, partly reflecting fears of trade sanctions and the negative effects of the then strengthening Japanese yen. Several assembly plants for Japanese automobiles were opened in the United States during the 1980s, two of them in Ohio (Reid 1995; MacKnight 1992). They formed the first phase of a relocation process that later attracted suppliers of component parts.

Japanese direct investment made significant contributions to economic growth in a number of Northwestern Ohio counties and municipalities. The city of Findlay (the seat of Hancock County) aggressively pursued foreign direct investment by creating a foreign trade zone and publicizing its proximity to Interstate 75 and small-town atmosphere. It became known in Japan as Friendly Findlay, an ideal location for Japanese transplant facilities. Hancock is one of the four study area counties that enjoyed increased manufacturing employment *and* decreased greenhouse gas emissions in the 1980s.

Thirteen of the twenty-one Japanese plants in the study area were greenfield investments: new manufacturing facilities in undeveloped areas. These newly constructed operations were usually state-of-the-art in terms of physical layout and production technology. Though energy- and emissions-efficient, however, those environmentally positive traits must be balanced against the role greenfield developments play in fostering urban sprawl and in increasing the consumption of fuel (and, therefore, augmenting greenhouse gas emissions) for transportation by household commuters and to meet the demands of just-in-time deliveries from suppliers and to the establishments that use the plants' products.

---

producing more greenhouse gases despite growing numbers of more fuel-efficient cars and recreational trucks. Moreover, industrial transportation emissions in the study area grew by nearly 20% in the 1980s, nearly one quarter of the regional transportation total. Industrial transport greenhouse gas emissions rose at higher rates in counties traversed by interstate highways, mostly due to the influx of Japanese automotive investment. To facilitate just-in-time delivery, Japanese automobile assembly plants and component parts works were sited adjacent to the region's interstates, particularly I-75.

Utilities

In 1990, utilities in Northwestern Ohio released a total of 4,475,000 metric (4,931,000 short) tons of carbon dioxide equivalents, or about 14% of total emissions from direct combustion of fossil fuels in the study area. Between 1970 and 1980, carbon dioxide emissions from the region's utilities increased by 9.5%, a trend reversed during the 1980s when study area power generators achieved a 4.2% decrease in carbon dioxide equivalents. Statewide, utility emissions increased by 3.7% from 1980 to 1990.

The difference between Ohio and study area utility emissions trajectories results largely from Toledo Edison company's growing reliance on nuclear electricity generation. Toledo Edison is an investor-owned company that provides electric service to approximately 300,000 customers in a 6,500 square kilometer (2,500 square mile) region of Northwestern Ohio. The area's two municipal and six cooperative facilities collectively serve only 15% of residential, commercial, and industrial customers in the region. Toledo Edison can generate 5,925 MW of electricity, of which 53% is coal-fired and 44% is nuclear-powered. In Ohio as a whole, 95% of all electricity is generated by burning bituminous coal. Since 1980, Toledo Edison has reduced its sulfur dioxide emissions by 62.3% and carbon dioxide emissions by 35.2%. Furthermore, it recycles more than 240 metric (265 short) tons of paper every year, saving about 4,500 trees and 6,812,000 liters (1.8 million gallons) of processing water. The American Electric/Ohio Power Company supplies electricity to more than 140,000 residential customers in the study region. American Electric recently embarked on *The Tidd Clean Coal Technology* project, a pioneering effort to establish the technical foundations for cleaner, more efficient coal-burning power plants for the next century. Project results will include lower operating costs coupled with substantial reductions in carbon dioxide emissions.

Agriculture

Although not as large a greenhouse gas emissions source as manufacturing and transportation, agriculture in Northwestern Ohio contributed comparatively high levels of emissions throughout the study period. This area produces 36% of Ohio's total agricultural output as measured by the market value of products sold, with 64% of total sales from cash crops and the balance from livestock sales. Agriculture's contribution to the region's gross local product (the value of all goods and services produced in Northwestern Ohio) declined from 2.7% to 2.0% between 1980 and 1990. From the perspective of the region's farm operators, per capita agricultural gross domestic product also declined by 3.3% during this period, owing to the interplay between a significant decrease in the number of farm operators, increasing scales of operation, and declining prices for agricultural commodities.

Agricultural components of the study area's greenhouse gas emissions were unchanged over the decade. The emissions come from four major sources: methane from manure management, methane from livestock, agricultural liming, and nitrous oxide from fertilizer application. In 1982, livestock and manure management accounted for most agricultural emissions. Nitrogen fertilizers and liming contributed significantly less to agricultural greenhouse gas emissions. Between 1982 and 1992, greenhouse gas emissions from livestock and manure management fell by 5.5%. Despite this decline, they remained the major sources of agricultural emissions in 1992, at 64.7% of the total. The contribution of nitrogen fertilizers to agricultural emissions increased absolutely and relatively during the decade. Nitrogen fertilizer emissions increased 10.0% from 1982 to 1992 and their share of agricultural greenhouse gas emissions rose from 25.5 to 29.6%.

Stable farming operations in Northwestern Ohio resulted in greenhouse gas emissions from agriculture that remained constant during the 1980s. Study area emissions from agriculture were 2,676,000 metric (2,948,500 short) tons of carbon dioxide equivalents in 1982 and 2,542,000 metric (2,801,000 short) tons in 1992. The decrease over the ten year period was less than 1%. Agriculture contributed 5.8% of Northwestern Ohio's greenhouse gas emissions in 1982 compared with 7.0% in 1992. Agriculture contributes disproportionately (7.0%) to Northwestern Ohio's emissions compared with its 2.0% share of the study area's gross local product.

Reducing emissions: technology and will

The emission reduction strategies and decisions available to Northwestern Ohio's local decision makers range from adaptive behavior on one hand to technological mitigation and abatement on the other. Global Change and Local Places surveys of major industrial emitters and key actors in the region suggest that solutions rooted in technology are preferred to those requiring changes in daily behavior.

Major emitters

Northwestern Ohio hosts four of the five most energy-intensive industries in the United States: petroleum and coal products; chemicals and allied products; primary metals; and stone, clay, and glass products (United States Bureau of the Census 1994a,b). These industries offer substantial potential for achieving energy efficiencies through incremental improvements and fundamental changes in production processes (United States Department of Energy 1995). From the perspective of local agency, a key challenge results from the locus of control: in many cases decisions to modify emissions at individual production sites must be made or approved in headquarters offices distant from the study area, where corporate officers may not favor or understand local commitments to reducing greenhouse gas emissions. Some owners of locality facilities, BPAmoco for example, have led international corporate efforts to achieve emissions reductions. The effects of that commitment on BPAmoco's local operations are impressive. BPAmoco has invested more than US$300 million to reduce pollution and greenhouse gas emissions at its installations and adopted such innovative corporate strategies as internal emissions trading (Box 5.3).

In other cases, an entire industry may define an emissions reduction plan, often in response to fears that government regulations will be imposed if the industry does not act first. One industry-wide plan that affects Northwestern Ohio is that formulated by the cement industry. Cement production emits one ton of carbon dioxide for every ton of cement produced. Reductions in cement plant emissions have been achieved in the study area through local operator participation in a broader industry plan.

Box 5.3 BP L ma Chemicals' internal emissions trading system

The geography of BPAmoco's global operations makes emissions trading an appropriate component of its corporate emissions reduction strategy. As part of a series of measures intended to mitigate greenhouse gas emissions, BPAmoco developed an internal emissions trading system in 1997. BP Lima Chemicals, located in Allen County, participates in BPAmoco's pilot project. The facility employs about 400 hands to produce acrylonitrile, an ingredient in synthetic fibers and plastics, and PCS nitrogen. BPAmoco initiated its program with an *allowance trading system*, which set a defined emissions cap for each facility. Facility managers may buy and sell emission credits depending on their future emissions needs and whether investments to accomplish abatement or buying emissions permits is cheaper. To facilitate this process, BPAmoco set the goal of reducing carbon dioxide emissions to 90% of their 1990 levels by 2010. This 10% reduction constitutes the baseline for determining emission caps for each participating facility or business unit and for allocating permits among them.

Each facility or unit received in 1999 an allocation of emissions permits valid through 2004. Each must restrict emissions to the quantities allowed by its allocated permits, in addition to any permits bought from other facilities or units. In the last 60 days of each calendar year, participants must demonstrate that they have permits sufficient for the year's emissions or face a monetary fine. If a facility will fall short, it may buy permits from other units with unused permits and thereby avoid the fine. Facilities or business units that have surpassed their emissions reductions targets may sell surplus emissions permits on the internal BPAmoco market or bank them for later use or sale.

The pilot emissions trading system is still at an early stage of development, requiring players such as BP Lima Chemicals to learn more about the scheme's complexities. Lima Chemicals is in the emissions trading market, mostly as a buyer inasmuch is its process improvements are scheduled for 2001 during its next plant shutdown. In recent years, BP Lima Chemicals has bought 15,425 metric (17,000 short) tons of emissions credits at US$18–23 per metric ton (US$20–25 per short ton).

Local industrial responses to global warming to date focus on technological solutions implemented by individual firms. A number of industries with facilities in the study area are pursuing initiatives directed toward using fuel more efficiently and reducing greenhouse gas emissions via waste reduction, energy saving, and production process improvements. Joint implementation initiatives also appear to be attractive to some of the area's major emitters. Overall, the key factors that motivate local industrial leaders to reduce emissions are corporate policy, education, sensitivity to the growing public awareness of the links between emissions and global climate change, technology transfer and workable market-driven solutions, carbon trading, and increased understanding of cross-sectoral linkages.

Government stakeholders

Northwestern Ohio's local governments portray the region as innovative and prosperous, and government officials view themselves as agents of change. Development agencies, such as the Toledo Port Authority, have acted aggressively to promote an image of the Toledo area as a vital region. With few exceptions, local government agencies consistently and persistently challenge any stereotypes that characterize the Toledo region as a part of the country's old Rust Belt. One example of the study area's wish to be viewed as economically vibrant, yet environmentally responsible, is Toledo's participation in the Cities for Climate Protection Program (Box 5.4).

Household perspectives

As industrial emitters reduce their contributions, Northwestern Ohio households contribute an increasing share of the region's total emissions. Household attitudes and behavior are difficult to understand given the complexities of degrees of belief that the threat of climate change is real, limited confidence in scientific knowledge and opinion, cultural dispositions and predispositions, and motivation. Individuals and households can certainly adapt to new beliefs and conditions, but the motivation to do so will most likely come from government and industry programs that offer households alternative choices and new options. To obtain some insight into householder attitudes and motivation, a survey was mailed to a random sample of 810 Northwestern Ohio households. A fourth of the surveys were completed and returned.

As is true for other major stakeholders, belief in the reality of global warming strongly conditions household responses, which can be clustered into three almost equal groups with respect to the credibility of global warming:

- those who perceive global warming caused by fossil fuel consumption to be a real problem (33%);
- those who are either indifferent to possible global warming or ambivalent about it (36%); and
- those who view global warming as unverified speculation (31%).

Respondents in rural areas are more skeptical of climatic change from greenhouse gas emissions than those who live in urban places. Estimates of the severity of climatic change in the next 50 to 100 years vary: 41% of respondents foresee no problem or a minor problem, and 30% judge future problems to be serious or extremely serious. When asked to rank the severity of various problems facing the United States, respondents ranked global warming last (11 of 11). Other environmental concerns, such as reducing air and water pollution and maintaining national parks for future generations, were considered more pressing (6 and 7 of 11, respectively). But global warming does outrank such other considerations as maintaining a strong military force, paying down the federal deficit, and reducing poverty and home-lessness. Questions phrased to arouse intergenerational feelings about global warming elicit somewhat different responses: when asked about possible global warming effects during

Box 5.4 Toledo's participation in the Cities for Climate Protection Campaign

Toledo's participation in the Cities for Climate Protection Campaign of the International Council for Local Environmental Initiatives exemplifies one local government's response to a common dilemma. Local politicians wish to foster pleasant, sustainable environments in their municipalities. At the same time, they take pains to be perceived as favoring development and job creation. When Toledo first considered joining the Cities for Climate Protection Campaign, competition was underway for a new US$1.2 billion Daimler Chrysler Corporation Jeep assembly facility to replace the aged Jeep assembly plant in central Toledo. The old plant employed 6,000 workers, and regional competition for the new plant was keen. Chrysler restricted potential new sites to within 80 km (50 miles) of the central Toledo plant. That placed Toledo in competition with smaller cities located in both Ohio and Michigan. Toledo and the State of Ohio offered Chrysler an incentive package totaling US$214 million to locate in Toledo. The package included tax credits, employee training programs, infrastructure grants, and environmental cleanups. The City of Toledo also agreed to buy 80 ha (200 acres) needed to expand an existing Chrysler facility and to bear the costs of relocating the residents and businesses that occupied the proposed building site.

Convincing city officials to join a visionary environmental program such as the Cities for Climate Protection Campaign while competition for the Chrysler Jeep plant was at its peak required ambitious efforts to allay fears that Cities for Climate Protection Campaign membership would be construed as anti-business. Those efforts included:

- ratification of a government resolution to complete an energy and emissions inventory;
- generation of a forecast of future emissions based on the inventory;
- development of a local action plan to achieve emission control targets; and
- initiation of an action plan composed of a series of policies to lead to the reduction of carbon dioxide and methane emissions.

Three factors helped convince Toledo's leaders to join the Cities for Climate Protection Campaign:

- the Global Change and Local Places project served as a conceptual catalyst. Its emissions inventories were well under way and good contacts had been established with municipal officials and local industrial leaders;
- key corporate stakeholders operating in the area (BPAmoco and Sun Oil, for example) endorsed Toledo's membership in the Cities for Climate Protection Campaign; and
- the mayor and other city officials were eager to promote the image of Toledo as a progressive city.

Winning the decision required several steps. First, local Global Change and Local Places team leaders contacted the City of Toledo's Office of Environmental Services about the Global Change and Local Places project and informed its administrators about Cities for Climate Protection Campaign opportunities. Once Environmental Services personnel were convinced that participation would benefit Toledo, they encouraged Global Change and Local Places researchers to submit a proposal to the International Council for Local Environmental Initiatives program. Toledo would qualify for an incentive payment of US$35,000 if it received an International Council for Local Environmental Initiatives grant. Toledo's linkages to the Global Change and Local Places project and the small number of applications to the International Council for Local Environmental Initiatives from old, Midwestern industrial cities made it an ideal candidate for International Council for Local Environmental Initiatives support. Selling the program to a somewhat skeptical City Council once the proposal was funded required two months of work. The Office of Environmental Services convinced the council that Cities for Climate Protection Campaign membership would benefit the city and local industries because:

- the program offered a way to strengthen ties between the city and the University of Toledo;
- increased energy efficiency would decrease greenhouse gas emissions while making local industries more cost-effective and competitive;
- financial incentives generated from the participation would support city programs such as the Municipal Energy Program started in 1996 and the Curbside Recycling Program;
- participation would not require doing anything that would hamper economic growth or that would not save money;
- participation did not require unquestioning belief in the existence of global warming;
- the grant would come from a private organization (International Council for Local Environmental Initiatives), not from the United States Environmental Protection Agency;
- when the grant money was spent, no further actions or obligations were required; and
- large companies with local operations such as BPAmoco, Sunoco, DuPont, and AT&T were already members of the Climate Wise program.

The Office of Environmental Services also emphasized that membership would provide the city with a number of specific benefits, including:

- free energy consulting services to Toledo and area businesses by Climate Wise;
- possibilities for additional grants in the future;
- showing citizens that environmental issues were important to city officials; and
- an improved working relationship with the United States Environmental Protection Agency.

Environmental Services personnel also pointed out that the University of Toledo Global Change and Local Places research team had already completed most of the emissions inventory.

The leaders of local industries were skeptical of the reality of global warming, but they also realized the public relations value of voluntary participation in the Cities for Climate Protection Campaign. The leadership of major firms such as BPAmoco and Sun Oil in endorsing the United States Environmental Protection Agency's Climate Wise program helped overcome fears that any information they provided to Environmental Protection Agency would be turned against them in the form of more stringent and costly regulation. A number of Office of Environmental Services workshops using local industrialists as leaders helped allay such apprehensions. The lesson from the Toledo experience is that implementing emissions reductions programs in localities requires the careful construction of coalitions of key stakeholders from as many sectors as possible. In this instance, University of Toledo Global Change and Local Places personnel were able to broker an effective partnership between government officials and industrial decision makers.

their lifetimes 54% of respondents thought *very little* would happen and none selected *great amount*. When queried about impacts during the lifetimes of their grandchildren, only 15% selected *very little* whereas 37% opted for the *great amount* response. Most Northwestern Ohio residents sampled regard local global warming as a phenomenon that will harm the area; 54% selected *bad* to characterize global warming's effects on the study area, compared to the 7% who thought warming might bring benefits. Plausible actions for reducing household greenhouse gas emissions focused on technology rather than behavior. Fuel switching or more efficient uses of energy were preferred to lowering thermostat settings or driving fewer miles.

Mitigation pathways

The responses of key industrial leaders, government officials, and householders suggest three plausible pathways to reduce greenhouse gas emissions in Northwestern Ohio:

- increased energy efficiency in production processes, a strategy easily promoted and adopted since it will benefit most industrial and commercial firms regardless of their attitudes toward global warming (energy efficiency reduces production and service costs);
- more fuel-efficient personal vehicles, a strategy that addresses one of Northwestern Ohio's increasing emissions sources in four ways:
  - smoother traffic flow to reduce congestion and dwell times at intersections, railroad crossings, and other bottlenecks;
  - improved vehicle maintenance through mandatory inspection programs to reduce emissions of volatile organic compounds, carbon dioxide and nitrous oxide;
  - reduced urban sprawl to curtail the number of vehicle miles traveled within the area; and
  - energy-efficient engines and vehicle designs to reduce transportation emissions.
- more aggressive urban planning for the residential and industrial sectors.

Table 5.4 Projected greenhouse gas emissions in Northwest Ohio, 2000–2020

Projection	Greenhouse Gas Emissions (metric tons)			
	1990	2000	2010	2020
Linear	36,534,000	30,005,000	23,617,000	17,228,000
Woods & Poole	36,436,000	40,297,000	43,006,000	45,867,000
Local Knowledge	36,436,000	42,268,000	45,153,000	48,145,000

A case study of improved urban design is the Jeep assembly plant's brownfield redevelopment, which focuses on central city redevelopment rather than a greenfield location that would promote urban sprawl. Brownfield redevelopment may help minimize costly inter-regional competition by reducing the scattering of parts and assembly facilities that presents large emissions bills for just-in-time component transport over long distances. Redevelopment also helps maintain core residential and commercial densities at levels that make viable communities sustainable.

Prospects

The Northwestern Ohio Global Change and Local Places research team expects the area's greenhouse gas emissions to rise at a relatively moderate rate in the next twenty years, consistent with the emergence of a new economic growth cycle. Efforts to contain these emissions should focus on rapid adoption of state-of-the-art innovations in energy-efficient technologies in industrial and residential settings. It appears unlikely that major behavioral changes to curtail greenhouse gas emissions will take place in the area in the near future. Other reduction strategies should aim toward an energy-efficient transportation system. Finally, incorporating land-use management and sustainable development planning into local programs would encourage brownfield redevelopment and slow urban sprawl.

Though encouraging, Northwestern Ohio's recent greenhouse gas reductions will not likely continue long into the future, whether projections are based on linear projections, a sectoral economic growth model, or local knowledge (Table 5.4). Assuming that recent trends will continue more or less linearly into the future yields emissions estimates for 2020 at a level half those of 1990, which is unlikely. Continued marginal gains in technological efficiency are not likely to be sustained for long, given the heavy investments already made in the 1990s. Expanding manufacturing, service, and transportation sectors is inconsistent with decreased emissions, despite the technological improvements that will doubtless be implemented. The scenario based on Woods & Poole (1998) medium growth forecasts for counties in the study area seems more realistic. It yields a 25% increase in emissions between 1990 and 2020, based on projections of moderate increases of 3.6% in agricultural output, 31.5% for industry and manufacturing, and 32.9% commercial and institutional growth. Combining the Woods & Poole scenario with the best judgment of the local Global Change and Local Places research team (the local knowledge projection) results in a slightly greater emissions increase (32%) than the Woods & Poole scenario's 25%. The adjustments that produce the difference are the Global Change and Local Places experts' opinion that some sectors will grow more rapidly than forecast in the Woods & Poole medium scenario. The estimate of 48,145,000 tons of carbon dioxide equivalent emissions in 2020 uses Woods & Poole high growth assumptions for the industrial and manufacturing, commercial and institutional, and production processes sectors, but Woods & Poole low estimates for the utilities and agricultural sectors. Those adjustments are based on the stability of agriculture in the study area, Toledo Edison's use of nuclear power, and its recent decision to observe the United States Department of Energy's Climate Challenge Participation Accord, which commits Toledo Edison to minimizing greenhouse gas emissions from its facilities.

REFERENCES

Ansari, A. and B. Modarress. 1987. The Potential Benefits of Just-in-time Manufacturing for U. S. Manufacturing. *Production and Inventory Management Journal*, **28**: 30–5.

Ballew, P. D. and R. H. Schnorbus. 1993. Auto industry restructuring and the Midwest economy. *Chicago Fed Letter*, **70**: 1–4.

Bergman, W. and W. Strauss. 1994. The Revival of the Rust Belt: Fleeting Fancy or Durable Good? *Chicago Fed Letter*, **80**: 1–3.

Bielen, M. 1998. Brownfields and Greenfields. *Newswaves*, **1** (1): 1–2.

Freeman, D. 1998. Personal Interview with D. Freeman, Toledo Metropolitan Area Council of Governments Transportation Planner, Toledo, OH. 16 December.

Gober, P. 1993. Americans on the Move. *Population Bulletin*, **48**.

Kenney, M. and R. Florida. 1991. How Japanese Industry is Rebuilding the Rust Belt. *Technology Review*, February–March: 25–33.

Klier, T. 1981, Assessing the Midwest Economy – A Longer View. *Chicago Fed Letter*, **107**: 1–3.

MacKnight, S. 1992. *Japan's Expanding U. S. Manufacturing Presence*. Washington: Japan Economic Institute.

Reid, N. 1995. Just-in-Time Inventory Control and the Economic Integration of Japanese-Owned Manufacturing Plants with the County, State, and National Economies of the United States. *Regional Studies*, **29** (4): 345–55.

Toledo Area Chamber of Commerce. 1991. *Office Space & Industrial Parks*. Toledo: The Toledo Area Chamber of Commerce.

United States Department of Commerce, Bureau of the Census. 1964. *Counties Employees Payroll and Establishments by Industry. County Business Patterns 1964: Ohio*. Washington, D.C.: United States Government Printing Office.

United States Department of Commerce, Bureau of the Census. 1981a. *1977 Census of Manufactures, Geographic Area Statistics, Part 2, General Summary*. Washington, D.C.: U.S. Government Printing Office.

United States Department of Commerce, Bureau of the Census. 1981b. *1977 Census of Manufactures, Subject Statistics*. Washington, D.C.: United States Government Printing Office.

United States Department of Commerce, Bureau of the Census. 1986. *Counties Employees Payroll and Establishments by Industry. County Business Patterns 1986: Ohio*. Washington, D.C.: United States Government Printing Office.

United States Department of Commerce, Bureau of the Census. 1985. *1982 Census of Manufactures, Geographic Area Series, Ohio*. Washington, D.C.: United States Government Printing Office.

United States Department of Commerce, Bureau of the Census. 1990. *1987 Census of Manufactures, Geographic Area Series, Ohio.* Washington, D.C.: United States Government Printing Office.

United States Department of Commerce, Bureau of the Census. 1994a. *Counties Employees Payroll and Establishments by Industry. County Business Patterns 1994: Ohio.* Washington, D.C.: United States Government Printing Office.

United States Department of Commerce, Bureau of the Census. 1994b. Manufacturing Energy Consumption Survey.

United States Department of Commerce, Bureau of the Census. 1996. *1992 Census of Manufactures, Geographic Area Series, Ohio.* Washington, D.C.: United States Government Printing Office.

United States Department of Commerce, Bureau of the Census. 1997a. *Annual Survey of Manufactures, Geographic Area Statistics.* Washington, D.C.: United States Government Printing Office.

United States Department of Commerce, Bureau of the Census. 1997b.*County Business Patterns 1995, Ohio.* Washington, D.C.: United States Government Printing Office.

United States Department of Energy. 1995.

United States Department of Energy. 1996. Climate Challenge Participation Accord.

Woods & Poole Economics, Inc. 1998. *County Projections to 2025: Complete Economic and Demographic Data and Projections, 1970 to 2025, for Every County, State, and Metropolitan Area in the U.S.* Washington, DC: Woods & Poole Economics. http://www.woodsandpoole.com/

6

Global change and Central Pennsylvania: local resources and impacts of mitigation

Andrea S. Denny, Brent Yarnal, Colin Polsky, and Steve Lachman

The Central Pennsylvania study area was incorporated into the Global Change and Local Places project because it provides telling contrasts with the other three study areas. Part of its economy relies heavily on local coal, a potential target for greenhouse gas mitigation strategies, and most of the Central Pennsylvania study area's greenhouse gas emissions originate in local coal resources. If forced to reduce emissions substantially, the area either would have to find ways to reduce sharply the emissions seemingly inherent in coal use, or stop using coal, removing one of the mainstays of the local economy. A second attraction for Global Change and Local Places was that the area has been a focus for research on regional development and climate change by faculty and students at the Pennsylvania State University, and attention to the effects of scale of analysis has been an integral part of that research.

Landscape, life, and livelihood

The five-county study area in the center of Pennsylvania (Figure 2.7) is a complex, sparsely populated amalgam of Appalachia and academia. Its population was 305,000 in 1990. For the study area in its entirely, population growth has been a modest 13% since 1970. The municipality of State College, home to Penn State University, is the largest settlement in the region with a population of 40,000 students and an equal number of permanent residents. The university population's influence is evident in the average age of residents in the respective counties. The mean for Centre County is 27 years, whereas the outlying counties' means range from 32 to 36 years. Centre County is the most urbanized part of the area, with 57% of its population so classified. The other four counties are 25% or less urban (Simkins 1995).

The five-county area is in many respects an integral part of Appalachia. The western counties are particularly economically depressed, which may account for the slow growth

The following Penn State faculty and students generously contributed their time to this chapter: Jimmy Adegoke, Marco Alcarez, Lauren Bloch, Dick Bord, Jeff Carmichael, Rob Crane, Bill Easterling, Ann Fisher, Gareth John, Rajnish Kamat, C. Gregory Knight, Steve Lachman, Shu-Yi Liao, Paul Mitchell, Rob Neff, Bob O'Connor, Becky Reifenstahl, Adam Rose, Chunsheng Shang, Robin Shudak, Damon Voorhees, and Nancy Wiefek. Primary support came from the National Science Foundation Human Dimensions of Global Change Grant SBR9521952, C. Gregory Knight, Principal Investigator.

Table 6.1 Employment structure in the
Central Pennsylvania study area, 1970–90

Values are percent of total employment.

Sector	1970	1980	1990
Manufacturing	42.2	34.6	25.0
Retailing	19.9	22.7	26.7
Services	15.8	20.1	26.0
Coal Mining	2.3	4.3	1.9
Other	19.8	18.3	20.4
Total	100.0	100.0	100.0

Table 6.2 Percentage employed in mining
in Clearfield County, Central Pennsylvania,
Pennsylvania, and the United States,
1970–90

Area	1970	1980	1990
Clearfield County	6.5	11.7	4.6
Study Area	2.3	4.3	1.6
Pennsylvania	0.7	0.9	0.4
United States	–	0.3	0.2

and population declines occurring in some areas (Simkins 1995). Despite the long-term growth and vigor of the Penn State community, many problems commonly associated with Appalachia persist in the study area. In 1990, 15% of the study area's residents had incomes below the poverty cutoff, and the average annual per capita income for the entire region was slightly less than US$15,000. Though low student incomes depress the study area average somewhat, pockets of genuine rural poverty are not uncommon. Early economic development in the area was based on extracting timber and metals and on farming. In recent years non-extractive industries and services have become more prominent components of the area's portfolio.

The study area's economy nests comfortably within those of Pennsylvania and the United States; manufacturing, retailing, and services constitute its three dominant sectors (Table 6.1). Manufacturing employment has decreased for more than 30 years, but manufacturing still employed a fourth of the area's labor force in 1995. The service sector increased from 15% of employment in 1964 to 28% in 1995, and retailing rose from 20 to 28% over the same period. Mining has remained more important locally than in Pennsylvania or the country, particularly in Clearfield County (Table 6.2).

Central Pennsylvania is conservative politically. Republican votes have dominated the last ten elections (Williams 1995). Religiously, conservative Protestants form the largest group. Enclaves of Amish and conservative Mennonites are scattered throughout the region,

Figure 6.1 Central Pennsylvania ridge and valley landscape.

particularly in the more agricultural areas (Zelinsky 1989, 1995). With the exception of the university population, the area is almost exclusively (97%) Caucasian (United States Bureau of the Census 1991).

The study area spreads across parts of the Appalachian Plateau and parts of Pennsylvania's Ridge and Valley region (Figure 6.1). Warped sedimentary strata underlie the surface, including large fields of bituminous coal. Coal underlies all of Clearfield County and portions of Clinton and Centre counties. The coals are part of the Lower Kittaning and Upper Freeport seams, both of which have high sulfur content. Less than 30% of the coal in Clearfield County and less than 50% of the coal in Clinton and Centre counties has been mined, leaving large reserves. There are also several small pockets of natural gas trapped within the coalfields (Marsh and Lewis 1995).

The study area has a humid continental climate, with precipitation distributed rather evenly through the year. There are noticeable variations in weather between the ridges and valleys, with the ridges having lower temperatures, stronger winds, and heavier precipitation than the valleys (Yarnal 1989, 1995). Severe weather in the form of thunderstorms, snow, and ice is common. The region is overcast, with an annual cloud cover of 68%. Cloud cover is more prevalent in winter because of the region's situation downwind from the Great Lakes. Conditions in the eastern counties are slightly more moderate than in the west. The average frost-free period is 160 days in Snyder and Union Counties compared with 140 days in Clearfield, Clinton, and Centre Counties (Yarnal 1989, 1995).

The study area's natural vegetation consists mainly of tall, broadleaf deciduous trees known as Appalachian Oak Forest. Cultivated crops include corn, wheat, and oats. Other

Table 6.3 Central Pennsylvania global warming
potential by gas

Values are percent of total GWP.

Gas	1990	1980	1970
Carbon Dioxide	95.25	95.07	94.96
Methane	4.33	4.46	4.58
Nitrous Oxide	0.42	0.47	0.46
Total	100.0	100.0	100.0

Table 6.4 Greenhouse gas emissions by industry in Central
Pennsylvania, 1970–90

Values are metric tons of carbon dioxide equivalent.

Sector	1970	1980	1990
Energy Use	12,219,000	13,961,000	16,404,000
Industrial Processes	487,000	490,000	250,000
Agriculture	299,000	363,000	401,000

agricultural production includes dairy farming and livestock, especially hogs and poultry. Cultivated areas are geographically intermittent because of variable soil quality, which depends in turn on the parent strata from which the soils have formed. The most extensive cultivated areas are located in Snyder and Union counties (Miller 1995).

Greenhouse gas emissions: carbon dioxide from coal burning

Estimates of greenhouse gas emissions were made for 1970, 1980, and 1990 for carbon dioxide, methane, and nitrous oxide. Emissions for 1990 totaled 17,459,000 metric (19,240,000 short) tons of carbon dioxide equivalent. Emissions for 1980 were 15,180,000 metric (16,728,000 short) tons and for 1970 13,338,000 metric (14,698,000 short) tons. Emissions increased by approximately 31% over the twenty-year period. In percentage terms, carbon dioxide dominated the emissions in all three years, followed in order by methane and nitrous oxide (Table 6.3).

Central Pennsylvania emission sources were broken down into four groups:

- energy (including stationary fuel use and transportation);
- industrial processes, which in the study area are dominated by the lime industry;
- agriculture (including manure management, enteric fermentation, and fertilizer use); and
- waste (including sewage, landfills, and human emissions).

Energy use is by far the largest source of emissions throughout the period from 1970 to 1990 in both absolute and relative terms (Tables 6.4 and 6.5). Consequently, much of the remainder of this chapter will focus on energy use in the study area.

Table 6.5 Percentage emissions by industrial sector in Central Pennsylvania, 1970–90

Sector	1970	1980	1990
Energy	91.6	92.0	94.0
Industrial Processes	3.7	3.2	1.4
Agriculture	2.2	2.4	2.3
Waste	2.5	2.4	2.3
Total	100.0	100.0	100.0

Table 6.6 Greenhouse gas emissions from coal in Central Pennsylvania, Pennsylvania, and the United States

Area	Percentage of Total Energy Emissions	Percentage of Total Emissions
Central Pennsylvania	82	77
Pennsylvania	46	40
United States	34	31

Driving greenhouse gas emissions: cheap, available coal

Because it so dominates local emissions, the energy sector is the key to understanding the forces driving Central Pennsylvania's greenhouse gas emissions. Transportation accounts for 11–12% of emissions from the energy sector. The remaining energy emissions come from stationary residential, commercial, industrial, and utility use of fuels including coal, oil, and natural gas. Coal is the primary fuel source in the region, due largely to the plentiful local supply. Approximately 82% of the energy emissions in the five counties are from the combustion of coal, while 77% of total greenhouse gas emissions in the five-county area originate in coal.

The high proportion of greenhouse gas emissions originating from coal use makes the study area distinct in comparison to Pennsylvania and the United States, which have more diverse emissions sources. Compared with Central Pennsylvania's 82% emissions from coal, only 46% of Pennsylvania's energy-related greenhouse gas emissions derive from coal and the comparable figure for the United States is 34% (Table 6.6). Furthermore, the local percentage of *total* emissions that can be attributed to coal burning is almost twice Pennsylvania's 40% and more than twice the United States proportion of 31% (Table 6.6).

Specific sources of coal emissions also vary across scales (Table 6.7). Approximately half the emissions from coal in the locality originate in electricity generation. Coal-generated electricity yields 35% of total greenhouse gas emissions in Central Pennsylvania, compared with 26% of total emissions in Pennsylvania and in the United States. The remaining coal combustion emissions arise in direct industry use (e.g. calcining lime) and in commercial and residential use, primarily for heating. The proportions of coal used for these activities

Table 6.7 Greenhouse gas emissions from coal use in Central Pennsylvania, Pennsylvania, and the United States

Area	Percent of Total Emissions from All Coal Uses	Percentage of Total Emissions from Coal Used to Generate Electricity	Percentage of Total Emission from Other Coal Uses
Central Pennsylvania	77	35	42
Pennsylvania	40	26	13
United States	31	26	5

Figure 6.2 Clearfield County bituminous strip mine in the Central Pennsylvania study area.

are exceptionally large in the study area (42% of total emissions) and small at other scales (13% of total Pennsylvania emissions and 5% of United States emissions). Abundant and inexpensive local coal in the five-county area accounts for these differences, which are evident in the landscape (Figure 6.2) and in emissions data in the form of lime production plants and the Pennsylvania State University's power plant.

Carbon dioxide from lime production (Alcaraz 1998)

In 1990, the two lime plants in the study area emitted nearly 250,000 metric (275,000 short) tons of carbon dioxide from the chemical reaction of calcination (the conversion of lime-stone to lime), or 1.50% of Central Pennsylvania's total emissions. In the United States as a whole, the lime industry accounts for only 0.25% of total emissions. When indirect emissions from the coal used to calcine the lime are added to direct lime industry emissions, the industry total increases to about 510,000 metric (562,000 short) tons of carbon dioxide, or about 3.0% of all study area emissions. The comparable total for the United States is 0.50% of total emissions. The disproportionate lime industry emissions reflect the presence of large, high-quality lime deposits in Centre County and the area's abundant bituminous coal deposits, making coal the preferred calcination fuel.

The many uses for lime and the absence of satisfactory substitutes suggest that such plants will continue to produce lime in similar or larger amounts in the future. Manufactured lime is needed for many industrial end uses, including the steel industry, utility power generation, pulp and paper processing, construction, and water and sewage treatment. Such environmental end uses as flue gas desulfurization and solid waste treatment are growing markets for lime. The enactment of Phase II of the Clean Air Act is expected to increase national demand for lime by 10–15%. Carbon taxes could affect the ability of local lime plants to operate profitably unless they were able to reduce greenhouse gas emissions. Unfortunately, calcining lime will always release carbon dioxide. Barring the installation of equipment to capture the carbon dioxide, these emissions cannot be avoided.

Interviews with Central Pennsylvania lime producers suggest that plants producing fewer than 180,000 metric (200,000 short) tons of lime per year would curtail production or close down in response to carbon dioxide regulation. Larger plants might be equipped with more efficient furnaces, switch to fuels that emit less carbon dioxide, or attempt carbon dioxide capture. Of the two lime plants within the study region, only one is large enough to consider emissions reductions. Switching to natural gas would seem the most likely way to achieve emissions reductions. Though the reductions resulting from fuel switching would be significant, the cost would be great. Potential reductions in the study area come to almost 113,000 metric (124,000 short) tons of carbon dioxide (22% of all lime industry emissions); natural gas is much more expensive than coal, especially in the study area. As of 2000, fuel costs were US$10.30 per metric ton of lime produced. Switching to natural gas then would have increased fuel costs by a third, to US$13.66 per metric ton of product. In terms of emissions, the cost to local plants would have been US$11.70 per metric ton of carbon dioxide not released as a result of using natural gas rather than coal, a figure that does not include any of the capital costs of converting to natural gas. Elsewhere in the United States, converting a lime plant to natural gas would cost about US$6.53 per ton of carbon dioxide avoided because of smaller differences between the cost of coal and the cost of natural gas in places where coal is not as abundant and inexpensive as it is in Central Pennsylvania. Switching to natural gas would cost more in the study area than elsewhere in the country, which could put the study area lime plants at a serious competitive disadvantage.

Greenhouse gas emissions at Penn State: the institutional view[1]

The Pennsylvania State University is situated at the geographic center of Pennsylvania. Founded in 1855 as an agricultural school, Penn State has grown into a major university with approximately 40,000 students at the University Park (State College PA) campus. The University continues to expand, and its growth and success are largely responsible for the continued growth of Center County. Because Penn State is one of the few major institutions encompassed by the Global Change and Local Places study, it offers examples of both the contributions a large public institution can make to a region's greenhouse gas emissions and the way such an institution might respond to its own responsibilities for greenhouse gas emissions.

Two ways to consider Penn State's greenhouse gas emissions are the institution's direct emissions and the emissions produced if Penn State is viewed as an end-user. The first approach is consistent with the inventory results presented in this chapter. The end-user analysis provides insights into mitigation ethics. The emissions produced directly by the university derive from electricity and heat generated by the campus steam plants, transportation on and around the campus, waste management, animal emissions from its farms, and byproducts of its lawn and athletic field management (refrigerants are excluded for consistency with the main inventory). In 1990, these sources contributed 195,000 metric (215,000 short) tons of carbon dioxide equivalent or 1.1% of the total emissions from the five-county region. The bulk of these emissions (91%) originate in energy use, making Penn State similar to the larger five-county study area. In addition to these direct emissions, Penn State is responsible for some emissions generated elsewhere. For example, the University purchases electricity to heat, cool, and light its buildings. Producing this electricity releases 182,000 metric (141,000 short) tons of carbon dioxide into the air each year. Students and staff who commute to Penn State generate greenhouse gases every day, while the university transports waste to landfills, which release additional emissions. When all end-uses are combined with the direct emissions, Penn State accounts for 338,000 metric (373,000 short) tons of greenhouse gases annually, or 1.9% of all study area emissions, a high proportion for a single point source, particularly because Penn State hosts no energy-intensive industries except for its experimental farms, which generate a negligible 1.4% of Penn State emissions.

Energy use on the other hand, largely in the form of electricity, yields 89% of Penn State's emissions, and coal is burned to produce most of the electricity produced on campus and purchased from commercial suppliers. In addition, limited public transportation and the geographic isolation of the Penn State campus cause intensive car use. Restricted on-campus housing and relatively low off-campus rents further contribute to the number of commuters. When considered on an end-user basis, transportation accounts for 6.4% of total emissions from the University.

Because Penn State is such a large producer of regional greenhouse gas emissions, it could affect the region's future mitigation options by making important contributions to emissions reductions. Unfortunately, greenhouse gas emission reductions do not now loom

---

[1] Based on Lachman (1999).

Table 6.8 Per capita emissions in Central
Pennsylvania, Pennsylvania, and the United States

Area	Metric Tons of Carbon Dioxide Equivalent per Person
Central Pennsylvania	57.3
Pennsylvania	26.0
United States	24.0

large in the university's plans. Function, aesthetics, and economics dominate recent Penn State planning documents. Reports contain brief allusions to energy efficiency, but largely as a function of building design. Penn State also plans to enlarge automobile parking areas, which will lead to increased vehicular emissions. Fortunately, some planned changes may help reduce future emissions. Campus bus services will be expanded and parking rates will be increased in order to reduce traffic congestion. Maintaining open spaces on campus and making new buildings more energy-efficient will also help.

Per capita emissions

The final consideration in comparing emissions at different scales is per capita emissions (Table 6.8). Per capita emissions from the five-county area are much higher than for the state and nation. The study area emitted 57.3 metric (63.1 short) tons of carbon dioxide equivalents for every 1990 resident. The state of Pennsylvania emitted 26.0 metric (29.4 short) tons for each citizen and the United States even less, with only 24.0 metric (25.9 short) tons per person. Central Pennsylvania's industry mix yields per capita emissions more than twice those of either Pennsylvania or the United States, a disparity driven by reliance on local coal for both electricity and direct energy generation. The disproportionate local use of coal is the source of greenhouse gases in Central Pennsylvania. The two coal-fired power plants within the bounds of the study area are the primary sources of emissions, while commercial, industrial, and residential coal furnaces are secondary sources.

Reducing greenhouse gas emissions: a region vulnerable to mitigation policy

Central Pennsylvania offers both an example of the challenges the United States faces in reducing greenhouse gas emissions without unacceptable economic impacts and also the potential for developing innovative technological and policy alternatives in order to meet such challenges. Penn State University provides an example of the way an informed end user could make its own decisions about its greenhouse gas emissions. Central Pennsylvania also demonstrates that sharp differences can exist among emissions profiles at different scales. The contrasts among Central Pennsylvania, the state of Pennsylvania, and the United States highlight the major differences that can exist among emissions profiles at different scales.

Furthermore, the greater difference between local and state profiles than between state and national profiles demonstrates that aggregating emissions data even to state levels may hide substantial local variability. Masking such differences could lead to serious mis-estimation of the local consequences of environmental legislation designed to reduce greenhouse gas emissions.

The Central Pennsylvania study area, for example, would be seriously affected by legislation aimed at reducing greenhouse gas emissions. Targeting the energy industry, particularly coal-powered utilities, would wreak local havoc. Coal-fired energy *is* likely to be a target of emissions reduction policies because it accounts for approximately one third of all United States greenhouse gas emissions. Furthermore, coal-fired energy currently produces more carbon dioxide per unit of energy than other fuels, making it an especially attractive policy target. Within the five-county region, greenhouse gas emissions from coal burning make up 77% of total emissions, suggesting that forced emissions cuts might have twice the impact locally as they would nationally. Forced fuel switching or emissions reductions could devastate local industries that rely on the inexpensive local coal as a fuel source. The region is additionally vulnerable because of its high per capita emissions. A comparatively sparse population and heavy industrial emissions yield high per capita emissions, although many of the products produced in the region – including electricity – are consumed outside its borders. Reductions mandated on a per capita basis could require study area residents to reduce greenhouse gas emissions twice as much as the United States average (Box 6.1).

Mitigation equity[2]

Most discussions of greenhouse gas mitigation ethics have focused on equity among countries, but in a country the size of the United States, with varied economic, physiographic, and cultural regions, equity is also an important issue. Differences in resources and infrastructure among regions may cause differential burdens to be placed on some regions. Establishing equitable mitigation schemes is important from an ethical perspective, but also in practical terms. Stakeholders are more likely to support actions they see as fair or even beneficial than those they believe to be unjust. Eleven possible mitigation paradigms and their potential impacts on the Central Pennsylvania study area are:

Horizontal

In a horizontal greenhouse gas reduction scheme, all regions would be forced to reduce emissions equally, regardless of proportionate emissions or past mitigation. This paradigm ignores differing impacts on local economies. In Central Pennsylvania, the horizontal paradigm would likely lead to further deterioration of the coal industry, improved efficiency measures, and lifestyle changes.

---

[2] Based on Rose and Stevens (1993).

Box 6.1 Potential impacts of mitigation: an analogy from Central Pennsylvania[3]
Andrea Denny and Brent Yarnal

Central Pennsylvania supplied coal to much of the nation throughout the late nine-teenth and early twentieth centuries. Coal consumption continues to grow in the United States, but Pennsylvania coal production declined in the late twentieth cen-tury in response to the Clean Air Act's restrictions on sulfur dioxide emissions from western Pennsylvania's high-sulfur coal (Munton 1998). Further cutbacks in coal use implemented to mitigate greenhouse gas emissions could trigger further production decreases. Mitigating greenhouse gas emissions by enacting blanket policies reducing coal use could seriously harm coal-producing regions (Glantz 1991).

Central Pennsylvania somewhat lags the national economy, in that manufacturing employment is declining more slowly and the service sector has not grown as rapidly locally as it has nationally, and coal mining continues to play a meaningful role in the local economy (Table 6.2). As much as 15% of the Clearfield County labor force was engaged in mining coal in the 1970s, and in 1993 (the last year for which data are available) mining made up 4–5% of local employment. The five-county study area continues to have a large stake in the health of the coal mining industry.

A carbon dioxide mitigation policy based on curtailing the use of coal could bring additional pressure to bear upon a region that is already economically depressed. The probable effects of such a policy are suggested by the consequences of the policies im-plemented to limit sulfur dioxide emissions to accomplish acid rain reduction. Sulfur dioxide emissions are limited via tradable permits, one of the strategies proposed to limit carbon dioxide emissions from fossil fuel combustion (Rose and Tietenberg 1993). Sulfur regulation began to affect the Central Pennsylvania study area beginning with the National Source Performance Standards (NSPS) regulations of 1979 and continuing with the 1990 amendments to the Clean Air Acts, which established the tradable permit system. Wyoming's low-sulfur coal is analogous to fossil fuels with lower carbon diox-ide emissions per unit of energy such as oil or methane. A shift-share analysis (Barff and Knight 1988) of competitive shifts for each year from 1980 to 1998 for Wyoming and Pennsylvania confirmed the hypothesis that Central Pennsylvania's declining min-ing employment and output was related to sulfur dioxide regulation (Figure 6.4). The analysis reveals differences attributable solely to the differences between Pennsylvania and Wyoming. After the introduction of sulfur dioxide regulation, coal mining jobs were lost in Pennsylvania in every year, whereas Wyoming experienced a competitive increase in jobs in nearly every year. Overall, Pennsylvania lost more than 10,000 min-ing jobs due to competitive differences while Wyoming gained over 4,000 jobs. Even when different costs based on mining productivity are taken into account, however,

---

[3] All data are from United States Department of Commerce, Bureau of the Census (1964–67 and 1969–95); United States Department of Energy, Energy Information Administration (1998); Oregon State University (1998). Modified from Denny (1999).

some of Pennsylvania's decline in mining employment is attributable to sulfur dioxide legislation, a conclusion supported by evidence from West Virginia. Production there is declining in the northern parts of the state where coal sulfur content is similar to Pennsylvania's coals, while production is increasing dramatically in the low-sulfur coal areas in southern West Virginia.

The results of sulfur dioxide mitigation suggest that legislation passed to limit carbon dioxide emissions from the combustion of coal, whether via tradable permits or via coal taxes (Muller 1996), would affect the national coal industry in ways similar to those seen in such high sulfur coal areas as Pennsylvania. In places like Clearfield County, those effects could be socioeconomically catastrophic. Although only 5% of the population is directly employed in coal mining, many others provide mining-related services or services to miners and their families. Although the overall regional economy might not be seriously affected (Abler *et al.* 2000), pockets of distress are likely to develop. Policy makers who wish to create equitable legislation must consider localities likely to be disproportionately affected by regulation. One possible solution is a domestic form of joint implementation (Harvey and Bush 1997) that would pair coal-producing areas with regions that generate no carbon dioxide from energy production, such as the Pacific Northwest.

Vertical

The vertical or Rawlsian approach would permit underprivileged regions to emit more greenhouse gases while prosperous regions faced emissions cuts. Such a policy would allow Central Pennsylvania (a relatively impoverished area) greater leeway in emissions, but such policy is unlikely in the current United States political climate.

Compensation

In this scheme, inefficient regions would be required to reduce emissions but would be compensated for their losses. The administration of such a strategy, however, would be complicated and troubling. For example, Central Pennsylvania could be compensated for being forced to reduce electricity production, but then all the surrounding regions that purchase that electricity would also be affected.

Sovereignty

This approach emphasizes regional control. All regions would be required to reduce emissions (although the required reductions could vary among regions), but each region would make its own decisions about how to meet its targets. This paradigm would allow Central Pennsylvania to reduce emissions in whatever manner it saw fit.

Egalitarianism

This plan assumes that all humans deserve the same opportunities, which in this instance would be achieved either by apportioning emissions based on population (in which case Central Pennsylvania would lose because of its sparse population and high emissions from coal) or by ensuring each region enjoys an equally satisfying lifestyle, a difficult and subjective criterion to meet.

Market

This approach assumes the adoption of a carbon tax, tradable permit system, or some other market-based method for deciding who is entitled to emit greenhouse gases. This type of system would probably affect Central Pennsylvania in the same way as the sulfur dioxide permitting system: that is, Central Pennsylvania would survive such a market system.

Lifestyle changes

Though unpopular, changing behavior will probably be necessary at some point in the future. However, such changes could be difficult in Central Pennsylvania, which is rural and dependent on automobiles and coal-generated energy.

The highly skewed proportion of local greenhouse gas emissions from coal burning makes the five-county study area highly vulnerable to broadly crafted legislation. The area could face an above-average burden if one-size-fits-all policy measures were applied to it, especially if they were based on per capita emissions. The region's disproportionate dependence on coal mining for employment (twice the national average in the five counties and eight to ten times the national average in Clearfield County), and on coal for fuel, could greatly multiply the local effects of forced emissions reductions. In a county where unemployment rates already far exceed the national level (9.5% locally as opposed to 5.4% for Pennsylvania and 5.6% for the United States), further job losses and economic instability would be difficult to accommodate (Oregon State University Government Information Sharing Project 1998).

Can coal be combusted differently?

Given that 77% of total greenhouse gas emissions in the five-county region are directly related to coal combustion, coal offers the greatest potential for emissions reduction, mainly through technologies that reduce emissions from coal-fired turbines in local power plants. Fuel switching and renewable energy sources are often seen as ways to reduce emissions from coal. Relying on these options, however, would severely damage an economy that relies on coal for employment and energy. Fuel switching would also be difficult and expensive in Central Pennsylvania because natural gas or petroleum would have to be piped in from elsewhere, an expensive undertaking given the region's rugged terrain and complex geology. Solar energy potential is limited by the area's heavy cloud cover. Nuclear power is unacceptable politically, given the region's proximity to Three Mile Island, the site of America's most infamous nuclear accident. Fortunately, several technological solutions promise to reduce emissions of carbon dioxide from coal. Adoption of enhanced technology appears to be one way the locality could maintain the status quo and meet emissions reductions that may be mandated in the future.

Emission reductions policies will likely focus on coal for two reasons. First, local utilities combust a great deal of coal and although they emit less than half the coal-derived greenhouse gas emissions in Central Pennsylvania, they are the largest single source of these emissions. Second, as noted above, approximately 45% of coal-related emissions in the study area come from just two coal-fired power plants. The remaining emissions are split among many small emitters in the commercial, residential, and industrial sectors. This distribution of emissions makes power plants attractive targets for policy changes. From a technological perspective, utilities are also the best choice for improvement. Currently, the most efficient commercially available coal-fired turbines operate at 39% efficiency, although efficiency is expected to rise over the next few years. Several proven technologies (supercritical heating, integrated gasification combined cycle, pressurized fluidized bed combustion, and indirectly fired cycles) could lower emissions even further.

The two corporations operating power plants in the study area participate in the Department of Energy's Climate Challenge Participation Accord, and they have taken steps to reduce greenhouse gas emissions (Figure 6.3). The continuation of such measures could play an important role in keeping emissions low in the future, and increased mitigation efforts will benefit both the environment and the financial performance of firms that undertake them. Deregulation of the utility industry in Pennsylvania has already forced corporations to improve efficiency, making it more likely that advanced coal technologies will be implemented as the existing furnaces age and are replaced.

Unfortunately, no single technological option exists that would reduce emissions to meet the guidelines established by the Kyoto Protocol. But available technologies do yield significant emission reductions and decrease the need for reductions in other sectors. If no action were taken within the coal industry and all other sectors maintained business as usual, study area emissions in 2012 would be 5,027,000 metric (5,540,000 short) tons of carbon dioxide equivalent above the Kyoto Protocol requirements. However, implementing either pressurized fluidized bed combustion or integrated gasification combined cycles would reduce the excess emissions to 3,078,000 metric (3,392,000 short) tons or 1,909,000 metric (2,104,000 short) tons of carbon dioxide equivalent, respectively. Although coal combustion contributes only 35% of total emissions within the study area, mitigation at power plants alone could realize 39–62% of the required emissions reductions, while generating increased profits for the power plants. Such large reductions from the utility industry would greatly reduce the amount that other study area industries need to mitigate in order to meet the Kyoto Protocol requirement of 7% below 1990 levels.

To assess future possibilities, three scenarios were constructed in accordance with the general Global Change and Local Places approach. First, a linear projection was calculated using data from the 1970, 1980, and 1990 inventories. This projection yielded

Figure 6.3 Central Pennsylvania study area coal-fired power plant.

total emissions of 19,243,000 metric (21,205,000 short) tons of carbon dioxide equivalent for 2000; 22,818,000 metric (25,145,000 short) tons for 2010; and 24,848,000 metric (27,383,000 short) tons in 2020 (Table 6.9). Linear projection produces an unrealistic view of future emissions because it does not consider changing economic and demographic patterns or changes in technology. A second set of projections based on data from the Woods &

Table 6.9 Central Pennsylvania emissions projections

Values are million metric tons.

Year	Linear	Woods & Poole	Modified Woods & Poole
1970	14,697,491	14,697,491	14,697,491
1980	16,727,620	16,727,620	16,727,620
1990	19,242,572	19,242,572	19,242,572
2000	20,787,878	21,354,612	20,690,019
2010	22,818,007	23,680,942	22,911,754
2020	24,848,136	25,640,860	25,074,166

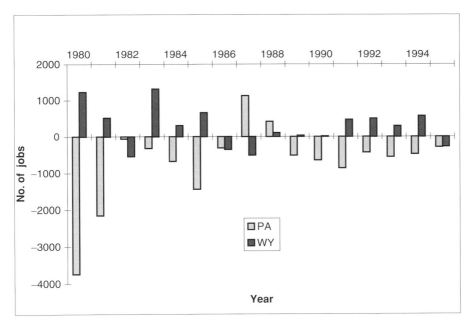

Figure 6.4 Shift share analysis of mining employment changes in Pennsylvania and Wyoming resulting from new sulfur emission regulations.

Poole (1998) State Profile for Pennsylvania and the United States Census of Manufactures yielded results similar to the linear projections. The close agreement between the two projections may result partly from the fact that portions of the 1970 and 1980 inventories were calculated by using ratios of economic activity that were also used to forecast emissions. In addition to the projection based directly on Woods & Poole data, a high-emissions scenario and a low-emissions scenario were calculated. The high-emissions scenario yielded results 20% higher than the standard projection, forecasting emissions of 23,530,000 metric (25,930,000 short) tons of carbon dioxide equivalent in 2000; 25,789,000 metric (28,420,000 short) tons in 2010; and 27,922,000 metric (30,770,000 short) tons in

2020. The low-emissions scenario results were 20% below the standard: emissions of 15,499,000 metric (17,080,000 short) tons in 2000; 17,187,000 metric (18,940,000 short) tons in 2010; and 18,612,000 metric (20,510,000 short) tons in 2020.

Although these projections incorporate foreseeable economic and demographic changes within the study area, several inherent errors mandate cautious use of their results:

- Some emitters, such as local power utilities, have begun to reduce total greenhouse gas emissions and their actions and plans are not considered in the projections;
- Future emissions from the lime industry (a major source in Central Pennsylvania) are based on the predicted economic activity of all industries, not just forecast activity in the lime industry. Fuel switching and increased kiln efficiency within the lime industry could reduce emissions.

A third set of projections attempted to incorporate these factors. Emissions from the utility sector for 2000 were set equal to 1990 emissions based on information from interviews with local utility-company representatives. Utility emissions for 2010 and 2020 were calculated in the same manner as the standard Woods & Poole projections, but with 2000 rather than 1990 as the base year. Emissions from lime processing were based on patterns of value added from lime production only, not from all manufacturing. The potential for reductions in emissions from the lime manufacturing industry were not included because of uncertainty in whether fuel switching from coal to natural gas would take place. All other categories were calculated in the manner described previously. These modified projection scenarios were in all cases lower than the original projections: 20,690,000 metric (22,800,000 short) tons in 2000; 22,912,000 metric (25,249,000 short) tons in 2010; and 25,074,000 metric (27,632,000 short) tons in 2020 (Table 6.9).

Prospects

Central Pennsylvania relies heavily on coal mining and coal use. Most residents of the region do not depend on coal for their livelihoods, but a significant number do, either directly or indirectly, and all residents depend on coal for energy. Given that dependence, the dominance of greenhouse gas emissions by carbon dioxide from coal combustion occasions little surprise. On the positive side, improving coal-burning technologies might make it possible to reduce the region's carbon dioxide emissions considerably. Reducing carbon dioxide emissions from coal combustion will be a necessary part of the United States effort to reduce overall greenhouse gas emissions. Even if technological innovations are adopted, residents of the five-county region will feel the effects of mitigation. Decreased demand for coal will hit local mining communities hard. Industries that find it impossible to operate without cheap coal will move out of the area or shut down altogether. The overall regional economy may not suffer greatly from such changes, but pockets of distress are almost certain to be evident where local economies are most directly and heavily dependent on coal. Because much of the coal burned in Central Pennsylvania produces energy and materials consumed outside the study area, judicious policies would devise measures to share the burdens of mitigation more equitably between local producers and distant consumers.

REFERENCES

Abler, D., J. Shortly, A. Rose, and G. Oladosu. 2000. Characterizing Regional Economic Impacts and Responses to Climate Change. *Global and Planetary Change*, **25**: 67–81.

Alcaraz, M. 1998. The Effects of Potential Carbon Dioxide Emissions Regulations on the United States Lime Industry at the National and Regional Scales. M.S. Thesis in Geography, The Pennsylvania State University.

Banff, R. A., and Knight, P. L. 1988. Dynamic Shift Share Analysis. *Growth and Change*, **19**, 1–10.

Denny, A. S. 1999. Greenhouse Gas Emissions from Coal Combustion in Central Pennsylvania: Addressing Vulnerability Through Technological Options. Unpublished M. S. thesis, Department of Geography, The Pennsylvania State University, University Park, PA.

Glantz, M. H. 1991. The Use of Analogies in Forecasting Ecological and Societal Responses to Global Warming. *Environment*, **33** (5): 10–15, 27–33.

Harvey, L. D. D. and E. J. Bush. 1997. Joint Implementation: an Effective Strategy for Combating Global Warming? *Environment*, **39** (8): 14–20, 36–44.

Lachman, S. 1999. A Greenhouse Gas Inventory of the University Park Campus of the Pennsylvania State University. M.S. Thesis in Geography, The Pennsylvania State University.

Marsh, B., and P. F. Lewis. 1995. Landforms and Human Habitat. In E. W. Miller, ed. *A Geography of Pennsylvania*: 17–43. University Park, PA: The Pennsylvania State University Press.

Miller, E. W. 1995. Agriculture. In E. W. Miller, ed. *A Geography of Pennsylvania*: 183–202. University Park, PA: The Pennsylvania State University Press.

Muller, F. 1996. Mitigating Climate Change: The Case for Energy Taxes. *Environment*, **38** (2): 12–20, 36–43.

Munton, D. 1998. Dispelling the Myths of the Acid Rain Story. *Environment*, **40** (6): 4–7, 27–34.

Oregon State University Government Information Sharing Project. 1998. *USA Counties 1998*. Retrieved November 1998 at http://govinfo.library.orst.edu/usaco-stateis.html

Rose, A., and B. Stevens. 1993. Efficiency and Equity of Marketable Permits for CO_2 Emissions. *Resource and Energy Economics*, **15**: 117–46.

Rose, A., and T. Tietenberg. 1993. An International System of Tradable CO_2 Entitlements: Implications for Economic Development. *Journal of Environment and Development*, **2**: 1–36.

Simkins, P. D. 1995. Growth and Characteristics of Pennsylvania's Population. In *A Geography of Pennsylvania*: 87–112. University Park PA: The Pennsylvania State University Press.

United States Department of Commerce, Bureau of the Census. 1964–1967 and 1969–1995. *County Business Patterns, Pennsylvania*. Washington, D.C.: United States Government Printing Office.

United States Department of Commerce, Bureau of the Census. 1991.

United States Department of Commerce, Bureau of the Census. 1997c. *County Level Population*. Retrieved September 1997 at www.census.gov.

United States Department of Energy, Energy Information Administration. 1998. Energy Information Administration Website. http://www.eia.doe.gov/fuelcoal.html (retrieved November 1998).

United States Department of Energy. 1997. *Clean Coal Technology Demonstration Program*. Washington, D.C.: United States Department of Energy. Publication DOE/FE-0364.

Williams, A. V. 1995. Political Geography. In E. W. Miller, ed. *A Geography of Pennsylvania*: 154–164. University Park, PA: The Pennsylvania State University Press.

Woods & Poole Economics, Inc. 1998. *County Projections to 2025: Complete Economic and Demographic Data and Projections, 1970 to 2025, for Every County, State, and Metropolitan Area in the U.S.* Washington, D.C.: Woods & Poole Economics. http://woodsandpoole.com/

Yarnal, B. 1989. Climate. In D. J. Cuff, W. J. Young, E. K. Muller, W. Zelinsky, and R. F. Abler, eds. *The Atlas of Pennsylvania*: 26–30. Philadelphia: Temple University Press.

Yarnal, B. 1995. Climate. In E. W. Miller, ed. *A Geography of Pennsylvania*: 44–55. University Park, PA: The Pennsylvania State University Press.

Zelinsky, W. 1989. Religion. In D. J. Cuff, W. J. Young, E. K. Muller, W. Zelinsky, and R. F. Abler, eds. *The Atlas of Pennsylvania*: 91. Philadelphia: Temple University Press.

Zelinsky, W. 1995. Cultural Geography. In E. W. Miller, ed. *A Geography of Pennsylvania*: 132–53. University Park, PA: The Pennsylvania State University Press.

PART THREE

Beyond Kyoto I: greenhouse gas reduction in local places

7

Changing places and changing emissions: comparing local, state, and United States emissions

William E. Easterling, Colin Polsky, Douglas G. Goodin, Michael W. Mayfield,
William A. Muraco and Brent Yarnal

The greenhouse gas emission inventories that currently inform abatement policy discussions have been developed almost exclusively from national-scale data, leavened only rarely with state or provincial inventories. Yet much of the capacity to abate greenhouse gas emissions necessarily resides within local institutions and communities. Policy may be debated and established at national and global scales, but it can be implemented only primarily by local action. This chapter examines how much information is lost when greenhouse gas emissions are estimated only at national scales (the United States in this instance) rather than at state or local levels, as in the four Global Change and Local Places study areas. That information may be critical to linking global and national policies to local actors and behavior.

Comparison of differences in the composition of greenhouse gas emission sources at three nested scales (national, state, local) for the four Global Change and Local Places study sites reveals good agreement in the *by-gas* composition of greenhouse gas emissions among national, state, and local inventories. Considerable differences are evident, however, in the *by-source* composition of greenhouse gas emissions among national, state, and local inventories. Geographical sovereignty is evident with respect to the composition of emissions, but geographical sovereignty does not hold for the sources of those emissions, suggesting that continuous monitoring of state and local emissions sources is needed to track geographical and temporal deviations from national trends.

Fugitive emissions and global perspectives

Human-induced greenhouse gases, once released into the atmosphere, recognize no boundaries. A molecule of carbon dioxide emitted from a smelter in Irkutsk today may be over Anchorage two days hence. For all practical purposes, it will be indistinguishable from any other molecule of carbon dioxide in the atmosphere and untraceable to its point of emission. This fugitive, anonymous property of emissions entices analysts into thinking that *where* greenhouse gas emissions take place matters little in the calculus of costs and benefits to society of limiting emissions. It would seem to follow that abatement of the seamless, increasingly dense blanket of greenhouse gases demands global solutions rather than research and action at regional or local scales.

Indeed, past and current analyses of the social efficiency of alternative emission reduction strategies are performed routinely using national or even global emissions estimates

(Nordhaus 1992; Dowlatabadi and Morgan 1993; Manne and Richels 1998). Ignoring the regional and local components of national or global inventories implicitly assumes either that sub-national variations in greenhouse gas emission sources and magnitudes do not exist or that they are unimportant if they are recognized to exist. That assumption may or may not be valid, but it can be and was tested by the Global Change and Local Places research team by examining in detail the proportional mix of greenhouse gas emission sources. Those sources were found to vary widely from the national mix depending on location and gas, and the underlying causes of those variations appear to be critical to the formulation of workable abatement policies.

Why scales of emissions estimates matter

Given the global nature of greenhouse-gas-induced warming, why worry if equivalence exists in sub-national and national emissions sources? Kates and Torrie (1998) contend that action to abate greenhouse gas emissions is inherently the provenance of local institutions, even though policy debate tends to be global. They opine that national policies may be needed to encourage abatement, but that decisions to reduce emissions are inherently local, residing in individual businesses, households, and agencies. Broad national prescriptions for reducing emissions that fail to accommodate local and sub-national conditions are unlikely to work. To be effective, international and national policies must mesh with the social tastes, habits, goals, and preferences that are most coherent locally. The disconnection between broad national abatement policies and strategies for local implementation has become more salient as nations and regions come to grips with the national implications of the Kyoto agreements.

Common sense suggests that inefficiencies in implementing abatement policies will accumulate if policy makers and legislators mistakenly presume that sub-national emissions mixes are scaled-down versions of the national emissions mixes. Inter-regional fairness also comes into play. Some analysts (Rose *et al.* 1998) argue for regional equity, asserting that all regions have an equal right to pollute and be protected from pollution. An abatement policy based on equity would strive to reduce regional emissions in proportion to regional contributions to total emissions: an area emitting 10% of a nation's total emissions should be responsible for ten percent of any national emissions reduction target – no more or no less.

A national abatement policy that ignores sub-national mixes of emissions sources could inflict excessive social and economic costs on some regions while letting other regions escape their proper shares of such costs. A national policy that relies on reducing emissions mainly from the largest source categories, coal-fired electric utilities, for example, could impose heavy economic and social costs on states or localities that generate energy for adjacent non-producing states. The inverse could be equally troublesome. Some states or regions may emit large quantities of greenhouse gases, but those emissions originate in sources that are not prominent at the national level. Agricultural states or regions that emit large amounts of nitrous oxide, for example, might be tasked to reduce emissions much less than their national shares of total emissions would justify simply because agricultural emissions are small components of national totals. Either instance would strain seriously the concept of sovereignty and common-sense concepts of equity. Emissions inventories that ignore or mask local and regional variations in magnitudes and especially kinds of greenhouse gas emissions are likely to lead to serious violations of the principle of equity, and therefore to result in policies that are widely criticized and extensively evaded.

Global Change and Local Places greenhouse gas inventory methods

The Global Change and Local Places research team tabulated greenhouse gas emissions on both *by-gas* and *by-source* bases. The general Global Change and Local Places methods for estimating greenhouse gas emissions are described in Chapter 2 of this volume. The Global Change and Local Places approach is consistent with the international reporting standard established by the Intergovernmental Panel on Climate Change, and adopted by the United States Environmental Protection Agency for United States and state-level emissions inventory protocols (United States Environmental Protection Agency 1995a,b).

The greenhouse gases tracked by Global Change and Local Places include carbon dioxide, methane, and nitrous oxide (Box 2.1, p. 41), as well as certain members of the growing family of fluorinated compounds (chlorofluorocarbons, hydrofluorocarbons, perfluorocarbons, and sulfur hexafluoride). These fluorinated compounds are known as ozone-depleting compounds or their substitutes. Ozone-depleting compound emissions are not included in Intergovernmental Panel on Climate Change and Environmental Protection Agency inventory protocols because the 1987 *Montreal Protocol on Substances that Deplete the Ozone Layer* outlawed production of these chemicals. Accordingly ozone-depleting compound emissions are excluded from the final emission counts in all national, state and Global Change and Local Places inventories. They are of important local interest, but methods for allocating them locally do not exist (Chapter 2). Because of the uncertainties in the data regarding the number and nature of activities that emit fluorinated compounds as byproducts, these gases will not be considered here (Box 7.1).

Emissions data were tabulated by major source category for 1990 for 26 of the United States and for all four Global Change and Local Places study sites and their surrounding states. Source categories of greenhouse gas emissions reported in this study reflect the major categories used in United States emissions inventories:

- fossil fuel combustion;
- biomass combustion;
- production processes;
- agriculture;
- waste disposal; and
- land-use change and forestry.

A rich variety of potential sources (and sinks) of greenhouse gases contribute to these major source categories (Table 2.1, p. 39). United States data for 1990–1996 were summarized directly from a recent United States inventory (United States Environmental Protection Agency 1998). State emissions data come from a series of Environmental Protection

Box 7.1 Fluorinated compounds

In terms of global warming potential, the most potent greenhouse gases are synthetic, fluorinated compounds used for a variety of domestic and industrial purposes such as solvent cleaning, refrigeration, air conditioning, and sterilization, and in aerosols, fire extinguishers, foams, electrical insulation and aluminum production (United States Environmental Protection Agency 1998). These compounds can be divided into two groups: ozone-depleting compounds and their substitutes. Ozone-depleting compound emissions are not included in Intergovernmental Panel on Climate Change and Environmental Protection Agency inventory protocols because the 1987 Montreal Protocol on Substances that Deplete the Ozone Layer outlawed production of these chemicals. Accordingly, ozone-depleting compound emissions are excluded from the final emission counts in all national, state and Global Change and Local Places inventories.

Ozone-depleting compound substitutes are themselves extremely powerful greenhouse gases, with greenhouse warming potential values as much as three orders of magnitude greater than carbon dioxide, methane, or chlorofluorcarbon potentials (for example, the value for chlorofluorocarbon is about 12,500 times that of carbon dioxide), and their production and use is growing rapidly. Estimating ozone-depleting substitute emissions is difficult because no government agency has been charged with tracking the wide variety of sources from which they emerge. The United States Environmental Protection Agency therefore estimates ozone-depleting substitute emissions for the United States on the basis of production and consumption in activities related to the use of the substitutes. State and sub-state estimates of these emissions are not publicly available. For these reasons, Global Change and Local Places analysts excluded ozone-depleting substitutes from their inventories. The exclusion should not greatly decrease the completeness of Global Change and Local Places inventories, as ozone-depleting substitutes account for only about 2% of national greenhouse warming potential totals (United States Environmental Protection Agency 1998).

Agency-sponsored state studies (United States Environmental Protection Agency 1999). Historical emissions estimates for 1970 and 1980 were interpolated using available time series data when estimates for those years were unavailable (Appalachian State University 1996). Global Change and Local Places emissions data required extensive local research because key data such as fuel sources of energy are collected only on a statewide basis and not by county or other local unit.

Considerable effort was made by the respective study area teams to acquire data from comparable sources across the scales and sites. Comparability was often stymied by the absence of consistency in data sources between scales and among sites, which results in significant but unspecifiable levels of uncertainty in the inventory estimates. Emissions from land-use change and forestry, and for ozone-depleting compound substitutes, were difficult

Box 7.2 Data comparability across scales

To compare emissions estimates across scales accurately within the limits of current emissions inventory methods, two conditions must be met: (1) the methods of derivation and aggregation used to produce the estimates must be consistent across and within scales; and (2) estimates must be complete for each source category across and within scales. Less than consistent and complete representation of emissions corrupts proportional comparison even if representation is consistent and complete. In the Global Change and Local Places study, neither condition could be met for emissions from land-use change and forestry, and for ozone-depleting substitutes. Consequently, those two source categories are excluded from the cross-scale analyses in this chapter.

Greenhouse gas emissions from land use are determined in part by two uncertain components: carbon sequestration and biomass combustion. Biomass combustion is a separate source category, lending some confidence in estimates of its emissions, even though Intergovernmental Panel on Climate Change protocols place carbon sequestration and biomass combustion into a common category that is the sum of the two components because carbon sequestration resulting from reforestation and afforestation typically takes place where biomass combustion does not, and vice versa. Ozone-depleting substitutes simply cannot be measured locally and regionally by any extant means (see Text Box 7.1).

to compare, and these two source categories are excluded from the cross-scale analyses in this chapter. Though less than optimal, excluding those source categories makes it necessary to recompute proportional contributions at all scales, without the two categories of land-use change and ozone-depleting substitutes. The Global Change and Local Places team believes a great deal can be learned from cross-scale comparisons of source categories even in the absence of data for carbon sequestration and ozone-depleting substitutes, but readers should be mindful of these missing data (Box 7.2). Site-specific estimates of carbon sequestration and ozone-depleting substitutes are incorporated into Global Change and Local Places data bases and analyses where they are available.

For 1970 and 1980, critical data are unavailable for major emissions categories such as transportation and for industrial and commercial emissions. The Global Change and Local Places research team developed a simple method for estimating historical emissions based on levels of measurable activity in each of those categories. Most estimates of 1970 and 1980 greenhouse gas emissions were estimated by using a formula similar to:

$$\text{Emissions}_{1980} = \text{Emissions}_{1990} \times (\text{Variable}_{1980}/\text{Variable}_{1990}),$$

where the variable was a measurable quantity closely related to emissions. To calculate emissions from transportation, for example, 1990 emissions were used as a basis for estimating 1980 and 1970 emissions by adjusting 1990 emissions using changes in three variables for which 1980 and 1970 data were available: number of registered vehicles, average fuel efficiency, and the average vehicle miles traveled. The respective study area teams generated the data used to construct the emissions inventories.

Cross-scale comparison of major greenhouse gases, 1990

The inventory of United States greenhouse gas emissions provides a baseline against which to compare state and local emissions (United States Environmental Protection Agency 1995a,b). In 1999, approximately 90% of total (Box 7.3) equivalent emissions came from fossil fuel combustion (United States Environmental Protection Agency 1999). Waste disposal and agriculture nearly tied for a distant second (Table 7.1). Though carbon dioxide was the dominant greenhouse gas emitted by fossil fuel combustion, methane was the most important greenhouse gas emitted by waste disposal, and methane and nitrous oxide were the leading agricultural greenhouse gas emissions from livestock and fertilizer application, respectively. Clearly, no policy designed to abate greenhouse gas emissions in the United States can ignore fossil fuel combustion. The by-gas data tell a similar tale: carbon dioxide itself accounts for about 86% of all emissions on a carbon dioxide equivalent basis (Table 7.2). Methane and nitrous oxide account, respectively, for approximately 12 and 2% of the remaining national total.

Carbon dioxide dominates emissions at the national, state, and local levels, with some state and local variations around the national carbon dioxide emissions in certain instances (Table 7.3). At 92%, Ohio's state carbon dioxide emissions as a percentage of its total global warming potential were higher than the national average. The Northwestern Ohio Global Change and Local Places study area's proportion also exceeded the national average. The pattern of 1990 carbon dioxide emissions in the Northwestern North Carolina study area was similar to Northwestern Ohio's, with state carbon dioxide emissions approximating the national average and study area carbon dioxide emissions slightly higher than the

---

Box 7.3 After carbon dioxide is there anything else?

Emissions of carbon dioxide dominate United States greenhouse gas emissions with very little interannual variation (Table 7.7). At the state level, carbon dioxide emissions range from 30% of all emissions in Colorado to 95% of Delaware's contributions (Table 7.8). Methane emissions in 1990 averaged 14% across the states and varied from 4% in Delaware to 64% in Colorado, highlighting the different emissions profiles of state-level industrial and agricultural economies. Nitrous oxide emissions ranged from zero in several states to 30% in Mississippi, averaging 3% across the states examined. One could argue from these data that carbon dioxide is the predominant greenhouse gas in most states, although large variations in its relative importance are possible at the state scale.

Table 7.1 United States greenhouse gas emissions by sector, 1990

Excluding land-use change and forestry, and ozone-depleting compound categories.

Category	Percent of Total Emissions
Fossil fuel combustion	89
Biomass burning	1
Agriculture	5
Production processes	1
Waste disposal	4
Total	100

Table 7.2 United States greenhouse gas emissions by gas, 1990

Excluding land-use change and forestry, and ozone-depleting compound categories.

Category	Percent of Total Emissions
Carbon Dioxide (CO_2)	86
Methane (CH_4)	12
Nitrous Oxide (N_2O)	2
Total	100

Table 7.3 1990 Study area and United States emissions by gas

Figures are percentage of greenhouse warming potential, excluding ozone-depleting substitutes and the land-use and forestry sector. Totals may not sum to 100% because of independent rounding.

Gas	GCLP Study Area				State				
	SW KS	NW NC	NW OH	Central PA	KS	NC	OH	PA	US
Carbon Dioxide (CO_2)	74	93	87	95	81	89	92	87	86
Methane (CH_4)	23	5	11	5	17	9	7	9	12
Nitrous Oxide (N_2O)	3	1	3	<1	3	2	1	3	2

Source: Region and state data from Global Change and Local Places analyses. United States data from United States Environmental Protection Agency (1995a).

national average. Kansas carbon dioxide emissions were less than the national average, and the Southwestern Kansas study area's carbon dioxide emissions were substantially lower. Methane emissions were second to carbon dioxide in terms of greenhouse warming potential at all three scales, ranging from 5 to 23% of greenhouse warming potential. Methane emissions in Southwestern Kansas (23%) were nearly twice the national average

Table 7.4 1990 Study area emissions by sector

Figures are percentage of Greenhouse Warming Potential, excluding ozone-depleting substitutes and the land-use and forestry sector. Totals may not sum to 100% because of independent rounding.

Gas	GCLP Study Area				State				
	SW KS	NW NC	NW OH	Central PA	KS	NC	OH	PA	US
Fossil Fuel Combustion	74	74	80	96	82	78	91	90	89
Biomass Burning	0	20	1	na	0	11	1	<1	1
Agriculture	24	3	1	1	12	6	3	4	5
Production Processes	<1	0	7	1	<1	1	1	<1	1
Waste	2	3	5	1	6	4	4	5	4

Source: Region and state data from Global Change and Local Places analyses. United States data from United States Environmental Protection Agency 1995a.

of 12%. Nitrous oxide emissions vary little among scales and locations, even in the predominantly agricultural study area of Southwestern Kansas.

Cross-scale comparison of sources of 1990 greenhouse gas emissions

Fossil fuel consumption predominated as a source of greenhouse warming potential in 1990 for the United States, for individual states, and for the four Global Change and Local Places study areas (Table 7.4). Percentage contributions ranged from 74 to 96% among the four study areas. At state scale, fossil fuel combustion emissions were consistent with national percentages, but with Kansas and North Carolina significantly lower and Ohio and Pennsylvania slightly higher than the United States value. Kansas methane emissions from agriculture (12% of total state greenhouse warming potential compared with 5% for the United States) depress its relative fuel combustion emissions. North Carolina fossil fuel emissions are proportionally lower than national fossil fuel emissions because of intensive biomass combustion by the state's furniture industry.

As one would expect, the largest cross-scale differences in proportional contributions to total greenhouse warming potential are evident when national data are compared with study area estimates. Among salient dissimilarities are:

- *Notable variations among study areas*   Even though it was the largest source of local greenhouse warming potential in all four Global Change and Local Places study areas, fossil fuel combustion was lower in percentage terms in the Northwestern Ohio, Northwestern North Carolina and Southwestern Kansas study areas than for the United States (Table 7.4). On the other hand, fossil fuel combustion overwhelmed all other local sources in the Central Pennsylvania study area, yielding greenhouse warming potential per capita much larger there than for the entire United States.
- *Central Pennsylvania relies heavily on coal consumption*   Coal use (embedded in fossil fuel combustion in Table 7.4) alone generated 77% of the Central Pennsylvania study area's 1990 greenhouse warming potential. Coal burned to generate electricity contributed

35% of the study area's potential, with the remaining 42% resulting from all other locality uses of coal. Corresponding figures for the United States are 31 and 26%, respectively. On a per capita basis in 1990, the Central Pennsylvania study area emitted approximately 57 metric (63 short) tons of carbon dioxide per person from all coal uses, compared with 17 metric (18 short) tons for the United States as a whole. Coal-based industries are heavily concentrated in Central Pennsylvania compared with the rest of the country. An emissions reduction policy that targeted coal users would have a proportionately larger impact on the region's well-being than on the national economy.

• *Agriculture-related methane emissions in Southwestern Kansas*   In Southwestern Kansas, methane emissions from agriculture and related functions (especially livestock finishing in confined feedlots) account for 24% of all study area greenhouse warming potential (Table 7.4). Agriculture generates nearly five times more greenhouse warming potential per capita in Southwestern Kansas than it produces in the United States as a whole. In this instance, even the use of state estimates for Kansas in place of national data would lead to serious underestimation of agricultural emissions in Southwestern Kansas.

• *Industrial/manufacturing fossil fuel combustion in Northwestern Ohio*   Northwestern Ohio conjures images of heavy manufacturing. Indeed, the proportional contribution of carbon dioxide emissions from fossil fuel combustion in the industrial/manufacturing sector (embedded in fossil fuel combustion in Table 7.4) to Northwestern Ohio greenhouse warming potential was nearly twice the national average (32 versus 18%). Major industrial/manufacturing contributors in 1990 were, however, the increasing number of light manufacturing plants in suburban locations around Toledo. Emissions from study area agriculture contributed slightly more to local greenhouse warming potential than in the United States. Though not a major agricultural emitter compared to Midwestern locations, the Northwestern Ohio study area demonstrates that even in largely urban regions, agriculture can produce significant greenhouse gas emissions.

• *Biomass combustion in Northwestern North Carolina*   Global Change and Local Places research in Northwestern North Carolina shows that the widespread assumption that atmospheric carbon fluxes from biomass burning and land use sum to zero may obscure meaningful local processes. In Northwestern North Carolina, the carbon dioxide emitted by biomass combustion is *not* entirely balanced by forest regrowth within the area because a large part of the material combusted is imported into the region. Moreover, biomass combustion provides significant energy (heat and electricity) in the region, conservatively estimated at 10% of total energy used in the study area in 1990. Local enterprises and households combust wood scrap from the large local furniture industry for heat and even for electricity. This *renewable* energy source in many cases displaces an equivalent amount of *non-renewable* coal, oil, and natural gas, thereby preserving the non-renewable energy base. Viewed at locality scale in this instance, carbon dioxide emissions from biomass combustion represent a fuel source and should be treated as such. Consequently, carbon dioxide and nitrous oxide emissions from biomass combustion were one fifth of all greenhouse gases released in the study area in 1990, and 11% from all of North Carolina. A broad national carbon emissions reduction policy uninformed by this kind of local information seems unlikely to credit such local efforts to substitute renewable carbon-based fuels for non-renewable ones.

If all that was needed to formulate effective emissions policies was identification of the most important source of greenhouse warming potential at state and local levels (fossil fuel combustion), then national percentages would suffice. If, however, a full accounting of the sources of greenhouse warming potential across all scales is desirable for policy-making on the basis of equity principles, then national data may induce large errors, especially at locality scale. Had only national percentages been used to impute 1990 contributions of state and local sources to greenhouse warming potential, how far off would the results we have examined be?

In Southwestern Kansas, methane emissions are about twice as important as in the United States, whereas in Northwestern North Carolina methane is about half as important as in the entire country. For Kansas, national data would yield overestimates of carbon dioxide emissions and an underestimate of methane emissions; for North Carolina, the reverse would occur. Perhaps there are no great surprises in these findings, especially once they have been articulated. A modicum of careful thought would lead most people to expect more methane in Southwestern Kansas than in other parts of the country and less carbon dioxide in Northwestern North Carolina.

More to the point of effective policies and strategies for abating greenhouse warming, however, are the facts that such local variations have rarely been considered heretofore, that documentation of their magnitudes has been even more infrequent, and that the departures from state and national means observed in the four study areas were larger than members of the Global Change and Local Places research team originally expected.

United States and study area greenhouse warming potential trajectories, 1970–1990

The Global Change and Local Places research team estimated total carbon dioxide equivalent emissions for 1970 and 1980 on the basis of 1990 emissions and relevant sectoral multipliers. Analysts did not attempt to disaggregate emissions by greenhouse gas or by source category for those years, as their goal was to use the decadal data for rough comparisons of emissions trajectories among the four study areas and between the study areas and the United States over the 1970–1990 period. Per capita emissions estimates were prepared for 1970 and 1980 by dividing estimated emissions by the census population counts.

The trajectory comparison documents a profound decrease (21%) in emissions in the Northwestern Ohio study area over the twenty-year period at a time when national emissions were increasing gradually (Table 7.5). Almost half (46%) of the carbon dioxide generated in the region in 1980 was released by fossil fuel burning in manufacturing, compared with 31% in 1990. Economic restructuring in the region and the associated shift to a more service-oriented economy led to an overall reduction in greenhouse gas emissions. The Northwestern Ohio transition demonstrates that a 1° study area can provide texture and resolution sufficient to identify important geographic and temporal changes in emissions related to industrial and residential relocations.

The Southwestern Kansas emissions trajectory illustrates the sensitivity of local emission paths to single events. From 1970 to 1980, the Southwestern Kansas greenhouse warming

Table 7.5 Greenhouse gas emissions index trajectories for
GCLP study areas and the United States, 1970–90

Value for 1970 = 1.00.

	1970	1980	1990
Southwest Kansas	1.00	0.96	1.33
Northwest North Carolina	1.00	2.55	2.64
Northwest Ohio	1.00	0.86	0.74
Central Pennsylvania	1.00	1.14	1.31
United States	1.00	1.07	1.12

Table 7.6 Greenhouse gas emissions per capita for GCLP study
areas and the United States, 1970–90

Values are metric tons of carbon dioxide equivalent.

	1970	1980	1990
Southwest Kansas	73.8	63.0	74.5
Northwest North Carolina	10.7	23.2	22.2
Northwest Ohio	30.0	25.5	22.1
Central Pennsylvania	49.5	50.3	57.3
United States	18.8	18.0	16.7

potential declined slightly, and then rebounded and increased more rapidly than United States emissions from 1980 and 1990 (Table 7.5). The rebound was caused by the opening of a new coal-fired electric utility (the Sunflower Power Plant) in the region. The added carbon dioxide emissions from the single plant were enough to reverse a declining trend for the entire region (Box 3.3, p. 66).

Per capita greenhouse warming potential time series can provide useful proxies for trends in the intensity of emissions from localities. Aside from the unknown ozone-depleting compounds that are absent from Global Change and Local Places computations, greenhouse warming potential for the United States changed little from 1970 to 1990 despite continued population growth (Table 7.6). In other words, per capita emissions declined. In Southwestern Kansas, on the other hand, per capita emissions were four times those of the country as a whole in 1970 and they increased between 1980 and 1990. The form of the Southwestern Kansas per capita greenhouse warming potential trajectory is almost identical to the plot of raw greenhouse warming potential, but the study area emissions remained more than four times greater than the national average throughout the period. Such discrepancies strongly suggest that population cannot be viewed as a reliable proxy for emissions across all scales of analysis. Global integrated assessment models such as IMAGE 2.0 (Alcamo *et al.* 1994) that rely on population as a major determinant of energy use and greenhouse gas emissions will be of limited use for small areas. Applying a population-based model to Southwestern

Table 7.7 United States greenhouse gas emissions by gas, 1990–96

Values are percentages of total.

Year	Carbon Dioxide (CO_2)	Gas Methane (CH_4)	Nitrous Oxide (N_2O)
1990	83.7	10.5	5.7
1991	83.4	10.7	5.9
1992	83.4	10.6	6.0
1993	83.7	10.4	5.9
1994	83.4	10.4	6.2
1995	83.5	10.5	6.0
1996	83.9	10.2	5.9

Source: http://www.epa.gov/globalwarming

Kansas would have yielded substantial underestimation of the study area's greenhouse gas emissions for the 1970s and 1980s.

Places, emissions, and policies

From the Global Change and Local Places perspective, an emission reduction policy must pass the test of equity across all relevant scales in order to be effective and geographically fair, which in turn implies that emissions inventories that do not permit fine-grained identification of emissions magnitudes and sources will be of increasingly limited value in future environmental policy formulation. Scale matters in the estimation of greenhouse gas inventories, and it will matter even more to the degree that equity criteria are widely accepted. Equity cannot be maintained unless accurate by-gas and by-source greenhouse gas emission inventories can be conducted at all relevant scales.

From a practical standpoint, the by-gas composition of greenhouse warming potential is similar enough to national and state mixes to permit estimates of state or local blends from national averages. Rankings of greenhouse gases from largest to smallest contribution to greenhouse warming potential are usually consistent across scales and study sites: carbon dioxide dominates all other greenhouse gases, followed by methane and nitrous oxide. Hence, national emission inventories can pass the test of equity across scales on a by-gas basis. One testable implication of these similarities is that an inverse relationship should also hold: a by-gas estimate of the composition of United States greenhouse warming potential derived from a sample of localities or states should be as accurate as a full accounting across all localities or states.

The by-source composition of greenhouse warming potential yields mixed results across scales. On the one hand, fossil fuel combustion sources dominate all other sources at all scales and for all study sites (see Box 7.4). Were the United States to choose an emissions abatement policy that targeted only the largest source of greenhouse gas

Box 7.4 If you have fossil fuels do you have it all?

If the objective of greenhouse gas inventories is to find the simplest way to account for the greatest proportion of emissions, then tracking carbon dioxide from fossil fuel combustion suffices. Carbon dioxide from fossil fuel combustion generates at least 75% of total emissions generally and in some areas close to 95% (Table 7.4). If associated methane and nitrous oxide emissions are included, the proportion is slightly higher. Addressing this single category, therefore, will yield a good approximation of greenhouse warming potential at any scale, an approach that offers the added benefit that energy-related data are easily captured and relatively accurate and complete for most places.

If the objective of analyzing greenhouse gas emissions inventories is to identify cost-effective and equitable emission reduction opportunities, however, then all source categories must be addressed, at all scales. A locality may not exhibit the same dominance by carbon dioxide from fossil fuels as is suggested by national or even state emissions profiles (Table 7.4). It may be more cost-efficient and equitable for people in Southwestern Kansas and Northwestern North Carolina to reduce emissions from agriculture and biomass combustion than to accomplish the same reductions from fossil fuel use. Focusing exclusively on carbon dioxide from fossil fuel combustion will hide material deviations from national and local emissions profiles to which policy makers need to attend.

emissions, a national emissions inventory based on the four Global Change and Local Places study sites would provide reliable information across scales. On the other hand, the relative contribution of sources that are secondary to fossil fuel combustion (agriculture and biomass combustion particularly) varies widely across scales. Using national percentages to impute local agricultural emissions in Southwestern Kansas or biomass combustion emissions in Northwestern North Carolina would generate serious underestimates of emissions from those sectors, and the protocol would fail the test of equity. Similar inventories for some states would also fail the same test.

Detailed geographical analysis of emissions makes it possible to identify the effects of high-impact events or a particular structural characteristic of a local economy on regional emissions. Only fine-grained inventories can pick up the effects on greenhouse gas trajectories of such events as the addition of the Sunflower electric power production plant in Southwestern Kansas, the shift to light manufacturing in Northwestern Ohio, and the importance of coal production in Central Pennsylvania. Analysis of roughly estimated greenhouse gas emissions for 1970 and 1980 and the more reliable estimates for 1990 revealed that per capita greenhouse warming potential in sparsely populated Southwestern Kansas was nearly four times higher than the national average during the period, an insight that could not emerge from a top-down approach. Emissions can be large even when population density

Table 7.8 Percentage of state greenhouse gas emissions and relative
rank, 1990

Based on million metric tons of carbon dioxide equivalent. States containing Global Change and Local Places study areas are italicized.

State	CO_2	CH_4	N_2O	Total	Rank
California	91	9	1	100	1
Pennsylvania	87	9	3	100	2
New York	83	17	0	100	3
Illinois	83	16	2	100	4
Georgia	91	8	1	100	5
North Carolina	89	9	1	100	6
New Jersey	91	9	0	100	7
Virginia	76	24	0	100	8
Alabama	76	24	1	100	9
Kentucky	79	21	1	100	10
Missouri	87	11	2	100	11
Tennessee	89	10	1	100	12
Wisconsin	86	12	2	100	13
Mississippi	55	14	30	100	14
Minnesota	84	12	4	100	15
Kansas	85	10	4	100	16
Maryland	89	5	6	100	17
Iowa	70	22	9	100	18
Utah	86	13	1	100	19
New Mexico	84	16	0	100	20
Washington	83	15	1	100	21
Oregon	91	8	0	100	22
Maine	93	7	0	100	23
Delaware	95	4	1	100	24
Hawaii	90	9	1	100	25
Montana	72	24	4	100	26
Colorado	30	64	6	100	27
Vermont	88	12	0	100	28

Source: United States Environmental Protection Agency (1999).

is sparse, suggesting that population can be used as a reliable predictor of emissions only
for very large areas.

United States national emission inventories deserve high marks for meeting the requirements of equity across scales for by-gas estimates. The same inventories are only marginally
reliable for producing accurate by-source estimates of greenhouse warming potentials across
scales. Regular compilations of state and local emissions source inventories appear to be a
prerequisite to tracking the spatial and temporal deviations from national emissions trends
that must inform equitable emission reductions policies.

REFERENCES

Alcamo, G., J. Joseph, J. Kreileman, J. S. Krol, and G. Zuidema. 1994. Modeling the Global Society-Biosphere-Climate System: Part 1: Model Description and Testing. *Water, Air and Soil Pollution*, **76**: 1–35.

Appalachian State University. 1996. *The North Carolina Greenhouse Gas Emissions Inventory for 1990*. Boone, NC: Appalachian State University Department of Geography and Planning.

Dowlatabadi, H. and M. G. Morgan. 1993. Integrated Assessment of Climate Change. *Science*, **259**: 183.

Kates, R. W., and R. Torrie. 1998. Global Change in Local Places, *Environment*, **40**: 39–41.

Manne, A., and R. Richels. 1998. On Stabilizing CO_2 Concentrations – Cost Effective Emission Reduction Strategies. *Environmental Modeling and Assessment*, **2**: 251–65.

Nordhaus, W. D. 1992. An Optimal Transition Path for Controlling Greenhouse Gases, *Science*, **258**: 1315–19.

Rose, A., B. Stevens, J. Edmonds, and M. Wise. 1998. International Equity and Differentiation in Global Warming Policy: An Application to Tradeable Emission Permits. *Environmental and Resource Economics*, **12**: 25–51.

United States Environmental Protection Agency. 1995a. *Inventory of U. S. Greenhouse Gas Emissions and Sinks: 1990-1994*. Washington, D.C.: U.S. Environmental Protection Agency, Office of Policy, Planning, and Evaluation. EPA-230-R-76-006.

United States Environmental Protection Agency. 1995b. *State Workbook; Methodologies for Estimating Greenhouse Gas Emissions, 2nd Edition*. Washington, D.C.: U.S. Environmental Protection Agency, Office of Policy, Planning, and Evaluation, State and Local Outreach Program. EPA-230-B-92-002.

United States Environmental Protection Agency. 1997. Pennsylvania Greenhouse Gas Emissions and Action Summary. State and Local Outreach Program, US Environmental Protection Agency. http://134.67.55.16:7777/DC/GHG.NSF/ReportLookup/PA

United States Environmental Protection Agency. 1998. *Inventory of U.S. Greenhouse Gas Emissions and Sinks: 1990–1996*. Washington, D.C.: United States Environmental Protection Agency, Office of Policy, Planning, and Evaluation. EPA 236-R-98-006.

United States Environmental Protection Agency. 1999. http://134.67.55.16:7777/dc/greenhouse_gas.nsf

8

Explaining Greenhouse Gas Emissions from Localities

David P. Angel, Samuel A. Aryeetey-Attoh, Jennifer DeHart,
David E. Kromm, and Stephen E. White

The Global Change and Local Places project case studies foster robust explanations of local, national, and international trends in greenhouse gas emissions. Knowing the specific events and processes responsible for changes in greenhouse gas emissions in particular places (sometimes called the *proximate* or *intermediate* forces of human-induced changes in the global environment) broadens understanding of possibilities for abatement or adaptation. Examples of proximate forces include the opening of a coal-fired power plant, or the growth of two-earner households and associated increases in automobile use. Proximate forces cannot be studied in isolation from the social processes that underlie them, the mechanisms and trends often called the *driving forces* of global change. Focusing on proximate forces and driving forces deepens understanding of greenhouse gas emission dynamics by emphasizing the degree to which the proximate forces that are so often the focus of policy responses are themselves determined by powerful social forces that policy makers often ignore.

The four Global Change and Local Places study areas were used as natural laboratories for teasing out details regarding the operations of proximate and driving forces. Case studies often provide contexts in which analysis and explanation are less refractory than they can be over larger areas (Box 8.1). Indeed, the distinctly different trajectories of greenhouse gas emissions for the four Global Change and Local Places study areas illustrate the ways emissions and changes in emissions over time vary in response to the different kinds of economic change that have occurred in the four areas. The impacts of population change and technology on emissions are different in different places. The decline in greenhouse gas emissions in Northwestern Ohio, for example, is closely tied to the restructuring of that study area's manufacturing establishments and their uneven success in lowering emissions per unit of output. In the Southwestern Kansas study area, changes in greenhouse gas emissions reflect the growth of feedlot animal farming, natural gas extraction, and the addition of a new coal-fired electricity-generating plant. Central Pennsylvania's greenhouse gas emissions are dominated by the heavy use of coal as an energy source. In the Northwestern North Carolina study area the growth of dispersed light manufacturing and attendant employment has driven local increases in greenhouse gas emissions. Abatement and mitigation strategies that are insensitive to local variability in emission sources are unlikely to be effective, and the Global Change and Local Places case studies are a good point of departure for identifying the different mixes of forces that operate at different geographical scales to produce local, state, national, and global greenhouse gas emission trajectories.

Box 8.1 Using local case studies to explain greenhouse gas emissions

Geographers and others generally put forward two interlinked arguments for using the local as an entry point for studying human–environment relations. First, many social processes operate partly at the local scale and are impacted by local conditions. Levels of political activism, for example, are often strongly linked to local circumstance. Attitudes and values are shaped in part by local experience. The local is also a scale of governance and of social regulation (Cox and Mair 1988). Of course processes and attendant outcomes rarely take place at a single scale; the norm is a complex multi-scalar dynamic that brings together processes operating at international, national, regional, and local scales (Swyngedouw 1997). The spatiality of social processes also changes over time, as exemplified by the impacts that new information technologies have had on the commodification of consumer culture around the world. The local entry point is likely to be of particular importance under those circumstances where outcomes are determined in part by processes operating at the local scale, and by the characteristics of particular places.

Second, local case studies can sometimes enhance the tractability of analysis and explanation. Disaggregation, whether by place, economic sector, or other dimension, is a strategy for simplifying the analysis of multiple, intersecting processes. The relevant question is whether the determining processes are structured in such a way that analysis is enhanced by disaggregation along geographical lines, and at the particular scale of the local. In many cases, the value added through local-scale analysis is likely to be limited, as for example, in studies of the impact of energy prices on energy consumption (where there is unlikely to be major variability in response at the local scale). But in other instances, such as the impact of local environmental conditions on attitudes toward greenhouse gas reduction policies, local case study analysis is a valuable research design. The contribution of local case studies is, in short, partly an empirical question linked to the spatiality of the processes under study.

Tracing greenhouse gases from sources to users

The Global Change and Local Places research team developed detailed inventories of emissions for the four study areas in order to identify the forces underlying changes in greenhouse gas emissions (Chapters 2 and 7). Analytically, the Global Change and Local Places project sought to decompose overall trends in greenhouse gas emissions into a series of material proximate forces, and then nest those proximate forces empirically and conceptually in their underlying social processes. Emission inventories can be quite sensitive to the boundaries of case study areas. A minor boundary change that excluded or included a major emission source such as a coal-fired power plant could substantially alter an area's greenhouse gas inventory. The emissions data created for the Global Change and Local Places project study areas are *direct* emissions from sources within selected regions. Global Change and Local

Places inventories do not include embodied or indirect emissions that are related to these sources but are not generated within the case study areas (Angel *et al.* 1998). For example, GCLP source totals include emissions generated by the use of an air conditioner, but not the (indirect) emissions created when the appliance was manufactured in a factory located outside the study area.

   This example highlights several issues. First, conducting full life-cycle analyses of the greenhouse gas emissions resulting from particular goods and services, such as Subak and Craighill's (1999) examination of the paper cycle, would add greatly to understanding the intricacies of global change at locality scale. Second, it is desirable where possible to allocate those emissions to their points of production or consumption. Some estimates of the amount of greenhouse gas emissions for which individual households are responsible, including both direct and indirect emissions, have been attempted (Morioka and Yoshida 1995; Gay and Proops 1993). Accomplishing such analyses is especially difficult for localities such as the four Global Change and Local Places study areas because of the high likelihood that the majority of emissions embodied in many goods and services will be generated outside the locality, and thus not included in standard emission inventories. Similarly, many of the goods manufactured in a locality or even a state will be shipped elsewhere, and the emissions released will not be charged back to the local inventory. Greenhouse gas emissions generated from natural gas compressor stations in Southwestern Kansas, for example, are allocated to the producing region rather than to the downstream users of that gas, or in the case of gas-fired power plants, to the eventual consumers of the electricity generated by burning the gas.

   Such detailed geographical accounting yields trenchant explanatory power and critical policy implications. Allocating both direct and indirect emissions to end-user households highlights the full greenhouse gas impacts of particular kinds of consumption, but such allocations also obscure the role firms play in the production of greenhouse gases and, consequently, potential opportunities for abatement at the site of production. Allocating emissions solely to points or areas of production downplays the impacts of consumption. Should the emissions originating in electricity generation, for example, be allocated to the power plant or to the end users of that electricity? Elsewhere we have argued that an important additional consideration is the locus of control over the processes involved. Alternatively, emissions could be apportioned among regions in proportion to the benefits accruing to the different areas from the production or consumption that give rise to the emissions (McEvoy *et al.* 1997). Or emissions could be allocated among regions according to their control over the processes that produce the emissions (Angel *et al.* 1998).

   In practice, both production-based and consumption-based inventories provide useful insights into greenhouse gas emissions. Where electricity is generated, for example, it is useful to know why a power plant is fueled by coal, gas, or another fuel, what determines the energy efficiency of the power plant, and other dimensions of the environmental performance of the facility. Within the Global Change and Local Places study areas, coal-fired power plants are the single largest point source producers of greenhouse gas emissions. In the Northwestern North Carolina study area, a single coal-fired power plant releases approximately one third of the region's total emissions. From 1970 to 1990, household consumption of electricity in that study area increased by 76%. To understand the forces driving greenhouse gas emissions, the processes and groups that use electricity must be identified, whether they are factories, farms, or households. In short, production and consumption are interdependent processes that should be investigated jointly to develop robust explanations of the dynamics of greenhouse gas emissions (Stern *et al.* 1997).

The greenhouse gas inventories presented thus far in this volume are organized primarily by source. Thus emissions from electric power generation are allocated to electric utilities, rather than to the people and places that use the electricity. In addition to these production-oriented inventories, Global Change and Local Places analysts have also calculated end-user (consumption) inventories for each study area. Those inventories are not full-life-cycle accounts that record all direct and indirect emissions from activities in the study areas; the original source-based inventories have been modified in two ways.

First, all direct emissions from sources have been allocated to one of three activity groups: commercial–industrial, households, or agriculture. Emissions from electric power production are also apportioned among the same three activity groups based on kilowatt-hour consumption. Similarly, transportation emissions were disaggregated based on three vehicle types: household passenger vehicles, commercial–industrial cars and trucks, and farm vehicles. These modifications link emissions more directly to end-user activities, an especially useful view of the dynamics of greenhouse gas emissions at the scale of the Global Change and Local Places study areas. Second, we include in our end-user (consumption) inventories the emissions resulting from electricity consumed in the study area only, and exclude from the inventories emissions produced by electricity generated within the study area but consumed outside its boundaries. Emissions from electricity generated outside the study area but consumed inside *are* included in the end-user inventories. In short, Global Change and Local Places counts emissions from electric power generation where the energy is used, not where it is produced (see Angel *et al.* 1998).

These adjustments highlight two facets of interest. First, Southwestern Kansas exports more electric power to surrounding areas than it imports from them. Second, the mix of fuels used to generate electricity varies from place to place. The Northwestern North Carolina study area contains a large coal-fired power plant, but a substantial proportion of the electricity consumed within the area is generated by a nuclear plant outside its boundaries. As a consequence, modifying the ways emissions from electricity generation are allocated may make a substantial difference in magnitudes of emissions for the Global Change and Local Places study areas, or other regions of similar size. For Southwestern Kansas, for example, total greenhouse gas emissions from *production* sources were 6.9 million metric (7.5 million short) tons in 1990, whereas the end-user inventory totaled only 5.4 million metric (6.0 million short) tons, a difference of 22%.

Empirical decomposition

The Global Change and Local Places greenhouse gas inventory modifications noted immediately above establish clear empirical links to the proximate forces that drive emissions, whether they be changes in the number of households in study areas or increased demand

Table 8.1 Study area end-user emissions by activity group, 1970–90

Measurements are short tons of carbon dioxide equivalent.

Study Area	Commercial–Industrial	Households	Agriculture	Total
Southwestern Kansas				
1990	2,775,467	1,393,236	1,903,108	6,071,811
1980	2,914,860	1,113,356	1,744,913	5,773,129
1970	3,851,148	942,722	1,147,800	5,941,670
Northwestern North Carolina				
1990	12,672,621	7,573,357	569,190	20,815,168
1980	13,119,005	7,996,075	610,062	21,725,143
1970	9,296,447	5,989,639	572,699	15,858,784
Northwestern Ohio				
1990	21,920,874	14,529,051	2,949,137	39,399,062
1980	27,737,965	14,845,051	2,898,064	45,481,080
1970	30,626,737	19,581,665	2,617,457	52,828,859

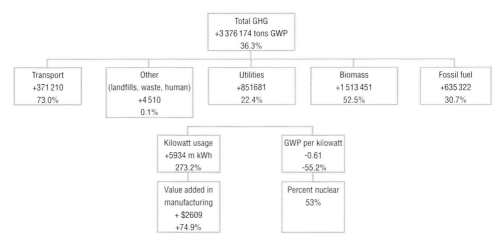

Figure 8.1 Change in greenhouse gas emissions from the industrial–commercial sector, 1970–90, in the Northwestern North Carolina study area. Measurements are in short tons GWP.

for electricity by local industry. Decomposing emission inventories by proximate forces yields further insights into patterns of change in greenhouse gas emissions. In Northwestern North Carolina, for example, emissions from the commercial–industrial sector increased by 3.09 million metric (3.4 million short) tons of carbon dioxide equivalent from 1970 to 1990, a rise of 36% (Table 8.1). Almost all the increase is attributable to increased energy consumption in the form of 0.77 million metric (0.85 million short) tons from electricity use, 0.54 million metric (0.64 million short) tons from fossil fuel combustion, and 1.40 million metric (1.51 million short) tons from biomass combustion (Figure 8.1). The significant contribution from biomass results from using furniture industry waste wood for fuel. The

Table 8.2 Drivers of change in greenhouse gas emissions within case study areas

Driving Force	Southwestern Kansas	Northwestern North Carolina	Northwestern Ohio	Central Pennslvania
Consumer-market demand	Demand for meat products	Exports of finished furniture	US market for automobiles	Market for coal
Regulation	National Energy Policy Act	Clean Air Act	Clean Air Act	Clean Air Act
Energy supply and price	Natural gas prices	Availability of natural gas	Oil supply and price	Coal prices
Economic organization	Growth of agro-food conglomerates	Diversified light manufacturing	Intensified competition in heavy industry	Importance of coal to the local economy
Social organization	Household structure	Changing household structure	Decline of 'family wage'	Household structure

large increase in electricity generation emissions is the product of increased use of electricity and is offset by increased consumption of electricity generated by nuclear power. Total electricity use by the commercial–industrial sector increased by 273% from 1970 to 1990, primarily as a consequence of manufacturing and service industry growth. Value added in manufacturing grew by 76% from 1970 to 1990 in constant dollars. Commercial–industrial emissions per unit of output actually declined over the twenty-year period, mirroring national trends. Overall, rapid industrial-led growth outweighed the gains achieved by using electricity generated by nuclear power and such other technology advances as greater energy efficiency in production processes.

Empirical decomposition of the forces underlying end-user emissions for other activity groups provides additional insights. Of the 1.4 million metric (1.58 million short) ton carbon dioxide equivalent increase (26.4%) in emissions from households in Northwestern North Carolina from 1970 to 1990, transportation contributed 0.65 million metric (0.72 million short) tons, documenting the key role of automobiles in emissions from households (Figure 8.2). Within the transportation sector, rapid growth (92%) in motor vehicle registrations from 1970 to 1990 was partly offset by a 55% improvement in automobile fuel efficiency. The increase in vehicle registrations in turn reflects larger households and an increase in the number of households, and a concurrent increase (18.6%) in average vehicles per household.

The end points of such empirical decomposition are not fixed (Table 8.2). One might also ask why the number of vehicles per household rose and cite such factors as the growth of two-earner households and the increasing affluence of many households. But that would beg the question about processes that have brought about the growth in two-earner households, and so on. Moreover, growth in the number of households in Northwestern North Carolina

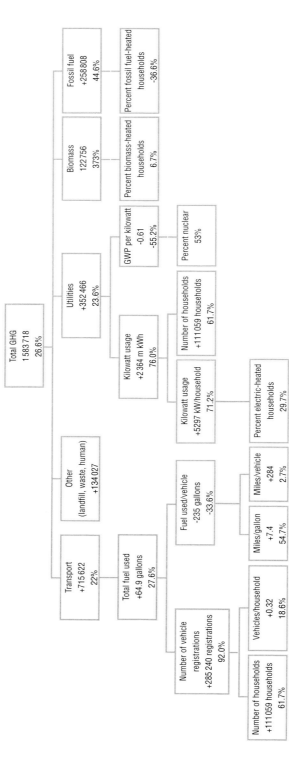

Figure 8.2 Change in greenhouse gas emissions from households, 1970–90, in the Northwestern North Carolina study area. Measurements are in short ttons GWP.

Table 8.3 End-user emissions for Northwestern North Carolina, 1970–90

Values are metric tons of carbon dioxide equivalent.

Sector	1970	1980	1990
Commercial–Industrial	9,296,000	13,119,000	12,673,000
Households	5,990,000	7,996,000	7,573,000
Agriculture	573,000	610,000	569,000
Total	15,859,000	21,725,000	20,815,000

Table 8.4 End-user emissions for Northwestern Ohio, 1970–90

Values are metric tons of carbon dioxide equivalent.

Sector	1970	1980	1990
Commercial–Industrial	30,627,000	27,738,000	21,921,000
Households	19,582,000	14,845,000	14,529,000
Agriculture	2,617,000	2,898,000	2,949,000
Total	52,826,000	45,481,000	39,399,000

is tied to the growth of employment and the array of processes underlying the transformation of the region into an area of light manufacturing. Rather than pursuing an exhaustive web of causal links, however, we might more profitably focus now on the major proximate forces of change affecting greenhouse gas emissions in the Global Change and Local Places study areas.

Major forces of change in greenhouse gas emissions

The empirical basis for explaining greenhouse gas emissions involves two related emission inventories, one organized by source (utilities, transportation, etc.), and the other organized by end users (commercial–industrial, residential, agriculture). End-user greenhouse gas emissions are largest in the Northwestern Ohio case study site as a result of high demand by heavy manufacturing enterprises for electricity and fossil fuels (Table 8.1). In all of the case study sites, commercial–industrial activity is responsible for the largest share of emissions, varying from 61% in Northwestern North Carolina to 46% in Southwestern Kansas. Households contribute 36% of all emissions in Northwestern North Carolina, 42% in Northwestern Ohio, and 23% in Southwestern Kansas. Agriculture is a major contributor in Southwestern Kansas, emitting 31% of the study area's emissions.

In Northwestern North Carolina both industrial and residential emissions rose rapidly from 1970 to 1980, but decreased by 4% from 1980 to 1990 despite continued economic growth within the region (Table 8.3). In Northwestern Ohio the trajectory of emissions is strongly downward, especially the commercial–industrial category, in response to declines in energy- and materials-intensive production in the steel and petroleum sectors (Table 8.4). Emissions in Southwestern Kansas declined modestly as natural gas extraction decreased

Table 8.5 End-user emissions for Southwestern Kansas, 1970–90

Values are metric tons of carbon dioxide equivalent.

Sector	1970	1980	1990
Commercial–Industrial	3,851,000	2,915,000	2,775,000
Households	943,000	1,113,000	1,393,000
Agriculture	1,148,000	1,745,000	1,903,000
Total	5,942,000	5,773,000	6,071,000

in the 1970s and then rose again as the gas industry recovered and as feedlot-based animal farming grew within the area (Table 8.5).

With emissions from utilities and the transportation sector allocated to end-user groups, the forces driving changes in greenhouse gas emissions in the Global Change and Local Places study areas can be identified easily. In all four areas, the dominant driver is the trajectory of local economic development and attendant changes in numbers of households. The specific mix of economic specialties differs for each Global Change and Local Places study area. Northwestern North Carolina has experienced rapid growth in light manufacturing, especially furniture manufacturing and food processing. That industrial growth in turn stimulated employment, in-migration of households, and the expansion of producer and consumer services. Northwestern Ohio's regional economy, traditionally dependent on such heavy industries as petroleum and steel, has become more competitive by focusing on automobile assembly and related industries. In Southwestern Kansas, agricultural expansion and innovation and the fluctuating fortunes of natural gas extraction govern the pattern of emission changes from 1970 to 1990. Greenhouse gas emission dynamics in each Global Change and Local Places study area are consequences of distinctive patterns of local economic development. Identifying those development dynamics is a prerequisite to understanding the dynamics of greenhouse gas emissions.

Additional insights into the forces underlying changes in emissions can be gleaned from production-based inventories. In Northwestern North Carolina, most of the decline in emissions from 1980 to 1990 occurred in electric utility emissions. Electricity consumption continued to increase, but a switch to nuclear generation and away from the initial dependence on coal-fired plants led to substantial reductions in greenhouse gas emissions from utilities. By 1990 half the electricity consumed within the region came from nuclear power plants. The resulting drop in emissions per unit of electricity generated shows how technology can affect the dynamics of greenhouse gas emissions. In practice, the effects of technology and the adoption of product and process technologies can be observed across all the end-user groups. Whether in the form of better automobile fuel efficiency, more widespread use of natural gas for home heating, or improved energy efficiency in manufacturing, technology helps determine levels of greenhouse gas emissions within all four Global Change and Local Places study areas. On balance, technology has reduced the greenhouse gas emissions per unit of economic output and consumption.

The Global Change and Local Places project adopted the $I = PAT$ (Impacts = Population × Affluence × Technology) identity as a framework for linking socioeconomic drivers to environmental change. Because population and affluence are easily measured, technology (T) is a residual, capturing the change in such environmental impacts as greenhouse gas emissions that are not attributable to changes in population and affluence. Operationally, the technology term melds the effects of such narrowly technical changes as better fuel efficiency with such behavioral variables as deciding to build a nuclear rather than a coal-fired electricity generating facility. Ultimately, the development and adoption of technology are themselves social processes. Such supposedly autonomous trends as decarbonization of production are socially determined, and that observation begs the question of what mix of social processes determines patterns of technology development, adoption, and use in the Global Change and Local Places study areas and at broader scales.

The dynamics of local economic development, changes in the numbers of households, and alterations in technologies in use underlie much of the change in greenhouse gas emissions that occurred in the Global Change and Local Places study areas from 1970 to 2000. As the case studies illustrate, however, a variety of other processes also come into play, ranging from decreases in average household size to increases in the average number of automobiles per household. A complete list of proximate forces that contribute to changes in emissions would be lengthy. The project's goal was to identify the processes that are empirically most important within each study area, in order to identify in turn the driving forces that produce changes in proximate forces.

Driving forces of greenhouse gases

The very idea of driving forces suggests identification of a set of ultimate determinants of changes in the proximate sources that govern the dynamics of greenhouse gas emissions. But most of the driving forces that are commonly identified, such as population growth and technology change, are themselves intermediate dynamics that are regulated by other social processes. Political economy and other forms of social theory suggest that an essential structure permeates social processes and nature–society relations. The Global Change and Local Places project's explanatory goals are more modest, focusing primarily on identifying the set of intermediate-level processes that have most influenced local economic development and technologies in use.

Different intermediate drivers of change have dominated changes in emissions in each Global Change and Local Places study area (Table 8.2). Those drivers are not end points in an explanatory chain, but they are some of the stronger links, given their influence on local economic development, associated increases in the number of households, technologies in use, and the resulting greenhouse gas emissions in the respective study areas. Five major groupings prevail: consumer/market demand, regulation, energy supply and price, economic organization, and social organization. No grouping is autonomous; all are linked to each other, and processes involved in each grouping may or may not be salient in a specific region.

As was expected, increased demand for the products and services produced in each area strongly influences local economic development and resulting emissions. Government regulation has also had substantial effects in the Global Change and Local Places study areas via the Clean Air Act, vehicle emission standards, and the National Energy Policy Act. At times regulation produces seemingly perverse results, such as the construction of a coal-fired power plant atop a large natural gas field in Southwestern Kansas. In general, though, legislation has fostered reductions in greenhouse gas emissions that otherwise might not have occurred. Energy supply and price also play key roles. Fluctuations in natural gas prices, for example, influenced the rate of gas extraction in the Southwestern Kansas study area. Elsewhere, the extension of natural gas pipelines to more homes in Northwestern North Carolina will reduce reliance on coal for home heating. Economic organization affects local fortunes and emissions as in the heightened international competition in heavy manufacturing industry that resulted in plant closures and the economic restructuring of the Northwestern Ohio study area on the one hand, and the emergence of large agricultural enterprises with the capital to expand feedlot-based animal farming in Southwestern Kansas on the other. Social organization in the form of changing patterns of work, of household structure, and norms and values also influence greenhouse gas emissions trajectories in all four study areas.

The five clusters of driving forces at work in the Global Change and Local Places study areas suggest three issues to be addressed in the interest of deeper understanding of the dynamics of greenhouse gas emissions. First, although I = PAT formulations sometimes yield reasonably accurate predictions of greenhouse gas emissions, even at locality scale the tripartite model fails to identify many of the key driving forces (DeHart and Soulé 2000). Its critical weakness is the relegation of technology to an all encompassing residual. The Global Change and Local Places case studies suggest that at a minimum, government regulation, economic organization, and social organization influence patterns of technologies in use in particular places at particular times, and technologies in use affect the overall trajectories of greenhouse gas emissions in regions the size of the Global Change and Local Places study areas.

Second, the majority of the driving force processes (Table 8.2) operate at scales beyond the local, and typically at national and international scales. Whether inter-firm competition in the steel and petroleum industries, national environmental and energy policies affecting the technology choices made by electric utilities, or the international markets for goods and services, development in localities is inextricably embedded in predominantly national and increasingly international processes. Local processes remain relevant, for localities are far from passive recipients of facilities, goods, and services determined by national and international dynamics. In Northwestern Ohio, for example, a local development coalition was a critical catalyst that fostered the social and political linkages necessary to attract new investment to the region. Moreover, greenhouse gas emissions are affected by local knowledge and experience, as is evident in the use of low-cost wood waste from furniture manufacturers in Northwestern North Carolina. Though mobilization of local knowledge may well become crucial to effective greenhouse gas abatement policies in the United States and elsewhere, broader forces operating at more extensive scales seem to be dominant.

Third, much of the authority over driving forces operating in the Global Change and Local Places study areas lies outside their boundaries. That observation occasions no surprise when applied to the national and international processes. One would not expect localities to shape national regulatory policies, the structure of interfirm competition, or the dynamics of technology development. But even at the scale of proximate forces, many local point-source greenhouse gas emitters are components of firms with headquarters outside of the study areas. The trend over time has been toward greater non-local control, as illustrated by the growing importance of multinational firms in each of the study areas. Local control over the processes that govern greenhouse gas emissions is limited at best.

Locality emissions

Local case studies can help explain the dynamics of greenhouse gas emissions in particular places. Because the mixes of proximate forces underlying changes in greenhouse gas emissions differ from place to place, case studies help to identify the key processes of change. Analysis of the four study areas upon which the Global Change and Local Places project focused has produced three broad insights concerning the proximate and the driving forces of greenhouse gas emissions.

First, most of the driving forces of change operate at national and international scales. National environmental regulations, the development of new process technologies, price-setting in energy markets, interfirm competition, and the energy efficiency of consumer durable goods, work at scales well beyond the local. The examples of local processes at work in the Global Change and Local Places study areas (growth coalitions in Northwestern Ohio and biomass fuels use in Northwestern North Carolina) were secondary in importance to broader forces. Though they cannot be ignored, the characteristics of particular places appear to shape the local outcomes of processes operating predominantly at larger geographical scales rather than altering those processes themselves.[1]

Second, effective control ('agency' in some literatures) most often resides outside localities. With few exceptions, the decisions that have shaped the emissions trajectories in the Global Change and Local Places study areas have been made elsewhere, as outcomes of national and international structural dynamics. National and international business strategies, facility investment options, fuel choices, technologies in use, and pollution control regulations, to name several factors, weigh more in the balance than local considerations. There exist indeed many opportunities for local actions to reduce greenhouse gas emissions, but the limited degree to which local actors shape locality greenhouse gas trajectories calls into question the durability of any reductions that might be achieved.

Third, even for localities, explanations are complex. The forces that drive greenhouse gas emissions cannot be reduced to a discrete, limited set of autonomous variables, such as population growth or increasing affluence. Such parsimonious explanations have an innate appeal and are sometimes supported by empirical associations between environmental impacts and aggregate measures of socioeconomic change, as in the oft-cited $I = PAT$ identity (Ehrlich and Holdren 1971). The Global Change and Local Places case studies

---

[1] As Massey (1994), Smith (1990), and others have argued, the characteristics of places that are experienced as 'given' at one point in time are themselves produced over time, including relevant physical characteristics, such as the availability of natural resources.

show that such variables as population growth (**P**), increasing affluence (**A**) and technology change (**T**) are themselves intermediate, with their own dynamics that are determined by multiple processes operating at a variety of geographical scales. Further, these underlying processes need to be understood not simply in terms of their empirical referents, but more fundamentally as a set of social relations with the natural environment.

Understanding the dynamics of greenhouse gas emissions in terms of both the proximate forces and driving forces of change forms the bedrock of effective policy responses. The analysis of the causes of emission dynamics in the Global Change and Local Places study areas has revealed the degree to which emissions are intertwined with local economic development. Mapping the particular mixes of processes that shape greenhouse gas emissions in localities will enable policy makers to develop flexible and place-sensitive responses to the challenge of greenhouse gas abatement. Achieving and maintaining flexibility will be necessary if policy makers are to balance the reduction of greenhouse gas emissions with such other societal goals as alleviating unemployment and improving socioeconomic welfare. Locally, nationally, and globally, the challenge to scientists and policy makers is to improve the energy and materials efficiency of daily life at rates that sufficiently offset the scale effects of economic growth.

REFERENCES

Angel, D. A., S. Attoh, D. Kromm, J. DeHart, R. Slocum, and S. White. 1998. The Drivers of Greenhouse Gas Emissions: What Do We Learn from Local Case Studies? *Local Environment*, **3** (3): 263–78.

Cox, K., and A. Mair. 1988. Locality and Community in the Politics of Local Economic Development. *Annals of the Association of American Geographers*, **78**: 115–35.

Ehrlich, P. R., and J. P. Holdren. 1971. Impact of Population Growth. *Science*, **171**: 1212–17.

DeHart J. L., and P. T. Soulé. 2000. Does I = PAT Work in Local Places? *The Professional Geographer*, **52**: 1–10.

Gay P., and J. L. R. Proops. 1993. Carbon-Dioxide Production by the UK Economy: An Input-Output Assessment. *Applied Energy*, **44**: 113–30.

Massey, D. 1994. *Space, Place and Gender*. Minneapolis: University of Minnesota Press.

McEvoy, D., D. C. Gibbs, and J. W. S. Longhurst. 1997. Assessing Carbon Flow at Local Scale. *Energy and Environment*, **8** (4): 297–311.

Morioka, T., and N. Yoshida. 1995. Comparison of Carbon Dioxide Emission Patterns Due to Consumer's Expenditures in UK and Japan. *Journal of Global Environment Engineering*, **1**: 59–78.

Smith, N. 1990. *Uneven Development: Nature, Capital, and the Production of Space*. New York: Basil Blackwell.

Stern, P. C., T. Dietz, V. W. Ruttan, R. H. Socolow, and J. L. Sweeney, eds. 1997. *Environmentally Significant Consumption: Research Directions*. Washington, D. C.: National Academy Press.

Subak S., and A. Craighill. 1999. The Contribution of the Paper Cycle to Global Warming. *Mitigation and Adaptation Strategies for Global Change*, **4**: 113–35.

Swyngedouw, E. 1997. Neither Local Nor Global. Globalization and the Politics of Scale. In K. Cox, ed. *Spaces of Globalization: Reasserting the Power of the Local:* 137–66. New York: Guilford Press.

9

Attitudes toward reducing greenhouse gas emissions from local places

Susan L. Cutter, Jerry T. Mitchell, Arleen A. Hill, Lisa M. B. Harrington, Sylvia-Linda Kaktins, William A. Muraco, Jennifer DeHart, Audrey Reynolds, and Robin Shudak

Greenhouse gas emissions arise from the acts of people in their local environments, and the sources of greenhouse gases and the driving forces behind their emissions are as varied as the localities included in the Global Change and Local Places project. These different mixes of emission sources and driving forces produce a range of local vulnerabilities, including different perceptions of the magnitude and nature of the problem, and different potential solutions. In order to reduce these emissions, a one-size-fits-all strategy may not work. Instead, analysts and policy makers may be faced with multi-faceted solutions that require an understanding of the dimensions of local vulnerabilities as well as local opportunities for prevention or reduction of emissions. These opportunities may not be realized, in large part because of differing perceptions of risks to different localities. Not only are there perceptual differences on the issue within and among economic sectors and governments, but there is also a significant degree of public indifference to the issue at local, state, and national levels.

Perceptions of climate change: thinking globally and mitigating locally?

Both lay and expert perceptions lead to different evaluations of the nature, extent, and scientific certainty regarding the existence or probability of climate change and its possible impacts (Redclift 1998; Stehr and von Storch 1995). These perceptions are often framed within highly localized sociocultural or sociopolitical contexts, which undoubtedly vary from place to place. In the same way, the willingness of local residents, industries, and businesses to reduce emissions may vary among places. People have variable wants and needs, and engage in a range of discourses at different scales, from the local to the global. These individual perceptions often lead to collective choices and the creation of social and political institutions to coordinate collective action. The role of human choice in risk decisions (which risks to bear and which ones require mitigation) will be reflected most noticeably at the local level, where people live, work, and play. It is at this level that the opportunities to undertake a range of greenhouse gas reduction strategies and the constraints on actually implementing those strategies may be the most pronounced and least understood. For that reason, analysts and policy makers need to know about the variability in local attitudes toward greenhouse gas emissions, mitigation, and adaptation strategies.

A second theme is the examination of local propensities to undertake adaptive and mitigative actions across the Global Change and Local Places study sites and local constraints affecting choices of actions. The social rationality of local decision-making on green-house gas emissions is a key focal point. Potential decisions by either households or industries regarding greenhouse gas emission reductions will be based on necessity and contingency, and made within the political–cultural milieux of the respective localities (Rayner and Malone 1998). While there may be some consensus on the nature of the problem, the results of Global Change and Local Places research suggest that broad-based adjustments may require localization in order to effect real progress in greenhouse gas reductions.

This chapter assesses local capacities for adaptive and mitigative actions by industrial sectors and social groups across the four Global Change and Local Places study areas. The analysis begins with the perspectives of major industrial emitters and assesses their awareness of climate change issues and the specific strategies and technologies that might be employed at their facilities to reduce greenhouse gas emissions. Then the opinions held by households – another stakeholder group – are reviewed in the context of prevailing national public opinion on climate change, as a window on broader societal viewpoints. The chapter concludes with an assessment of the mitigation potential of residents in our study areas based on a perceptual survey of their attitudes toward greenhouse gas emissions and their willingness to undertake actions to reduce them.

The view from the inside: major emitters

Given the rates and magnitudes of greenhouse gas emissions at the study area scale, what are the levels of knowledge of mitigative actions, experience with such actions, and alternatives to current practices that would encourage reductions in greenhouse gas emissions by the major emitters in the area? In the Global Change and Local Places study areas, the major emitters include energy providers (electric power generation installations, natural gas compressors, and universities), industries (manufacturing, oil companies), transportation, and agriculture (farming and feedlots). While local variations in the relative contributions of each sector to the total greenhouse gas warming potential exist, electric power generation was a dominant source of greenhouse gas emissions in all four study sites. In order to assess the perspectives of these major emitters, interviews were conducted with a representative of each major point source emitter in each study area (Chapter 2).

Four questions were posed during the course of open-ended interviews designed to:

- assess the level of facility concern and awareness of climate change;
- elicit specific strategies for decreasing greenhouse gas emissions at the facility;
- determine who felt responsibility for greenhouse gas mitigation activities in the local area; and
- identify specific technologies that could be used to reduce greenhouse gas emissions in each facility.

The interviews represent highly localized and time-dependent opinions. They offer snap-shots of attitudes that may not be generalized across all sites or industrial sectors, or even later time periods for these sectors and their sites.

Do you believe global warming to be a real concern?

As expected, considerable variation in levels of concern exists among the major emitters. Opinions range from little concern (Southwestern Kansas) to an implicitly high level of concern as demonstrated by participation by some Northwestern Ohio officials in the International Council for Local Environmental Initiatives Cities for Climate Protection Campaign.

Industries emitting the largest amounts of carbon dioxide were keenly aware of climate change issues and demonstrated considerable knowledge of potential regulatory action, especially the Kyoto Accord. Large power companies such as Duke Energy and Pennsylvania Power and Light participate in global warming conferences and engage in lobbying efforts at the corporate level. There was some concern expressed (particularly by natural gas and electric power industry representatives) that greater regulation of greenhouse gas emissions would harm specific industries without addressing underlying problems. This was especially true in Southwestern Kansas, where emitters believed that the United States Congress might over-react to global warming concerns and enact legislation based more on worst-case scenarios than on probable realities. In contrast, the major emitters in Toledo were working actively on local action plans (under the International Council for Local Environmental Initiatives' Climate Protection Campaign) that include analyses of appropriate abatement or adaptive strategies for reducing greenhouse gas emissions. Similarly, major emitters in Central Pennsylvania claimed that greenhouse gas emissions were an important component in their long-range environmental planning. Industries with lower emissions were less aware of climate change issues and potential regulations, and instead focused on air-quality regulations such as those embodied in the Clean Air Act, especially in Northwestern North Carolina among furniture manufacturers.

Most major emitters do not recognize emissions at the local level as contributions to an enhanced greenhouse effect. In Southwestern Kansas, for example, methane released by cattle is well recognized, but the link between that methane production and global climate change is not generally recognized. The same is true for landfills in Northwestern Ohio. In Southwestern Kansas, some interviewees pointed to the possible benefits of greenhouse gas emissions, such as enhanced crop growth due to increased carbon dioxide levels.

An apparent generational split is evident in levels of concern about climate change at the managerial level. Interviewees at all sites (especially in Southwestern Kansas) believed that the idea of human-induced climate change was more accepted by younger people, who were also judged to be more likely to be concerned, take climate change seriously, and take mitigative actions in response to their concerns. However, in some industries (such as natural gas) older and more resistant people tend to occupy decision-making positions and thus are in a position to thwart attempts at reductions by younger officials.

Specific actions for reducing greenhouse gas emissions

Most major emitters focused on two generic issues and approaches: technological fixes and economic incentives. In Northwestern Ohio and Central Pennsylvania, for example, the major emitter response emphasizes technological solutions implemented at the firm level, actions that managers hope will achieve greater energy efficiencies. Fuel switching was mentioned most often as an option in Northwestern North Carolina, while in Central Pennsylvania the movement from coal to nuclear power to generate electricity was mentioned. Despite knowing that switching from coal to natural gas in some of the thermal generating plants will reduce emissions, there was concern over the capital cost of such switches and over long-term operating costs. Some major emitters considered ceasing their own power production and purchasing power from the local utility. This would shift the emissions burden to another source, but not decrease it for the region. Overall, the best selling point for any mitigation action, according to the major emitters, is economic.

Some utilities participate in energy efficiency projects such as the Environmental Protection Agency's Energy Star program and Department Of Energy's Climate Challenge Accord program. Feedlots in Southwestern Kansas use wind turbines for some of their electricity needs. In the natural gas industry, new compressor motors reduce emissions and also make compliance with the Clean Air Act of 1990 easier. In the transportation sector, economic motivations seem paramount. Strategies such as increased efficiency in scheduling (thereby reducing fuel use) or the use of alcohol as a gasoline additive, were seen as options. In Northwestern Ohio, waste reduction, energy savings, carbon dioxide emissions trading, and product improvements topped the list of potential strategies. Emissions trading initiatives also appear to be a key consideration for many of the major emitters in the Northwestern Ohio study area.

Who should be most responsible for greenhouse gas reductions?

Major emitters believe that if emissions policies are to be developed, they should be implemented at state and regional scales. While they recognize the need for federal oversight, they trust state and local officials more than control at the federal level. Especially in Southwestern Kansas, emitters recognize a potential for local interests to become corruptible, making some oversight necessary, preferably at the state level. The federal government (through agencies like Environmental Protection Agency) is viewed quite negatively in at least one study area, being seen as capricious in policy development and implementation, with inconsistent year-to-year legislation. Unfunded mandates are an especially sensitive issue. In Northwestern Ohio, emitters believe that control should be exercised at the firm level for the optimal results. In Central Pennsylvania, deregulation seems to have increased greenhouse gas emissions by electricity producers, who must produce power cheaply in order to stay competitive. The tension between environmental considerations and price makes plant efficiency a more pressing consideration.

Emitters in all four study areas argued that the driving force behind the policy should be derivative of industry goals, not governmentally driven. Bottom-up or middle-down goal-setting was deemed the most desirable. Cooperation between local actors (including trade organizations) and government agencies was held to be more desirable than the imposition of regulations. Major emitters have little faith in behaviorally driven alternatives such as changes in public attitudes or consumption patterns, and are more likely to opt for increased efficiency and other technological solutions.

Specific technologies to reduce emissions

The Northwestern Ohio region was oriented more strongly toward technological methods of greenhouse gas emission reduction than any of our other three sites. Elsewhere, interviewees who worked in such specific industries as transportation or feedlots were relatively well informed about existing technologies that offer greater efficiencies. Feedlot operators, for example, mentioned the need for greater development of wind power and the strategy of covering manure lagoons (Harrington and Lu 2002). Electrical utilities suggested that carbon dioxide trapping may become important in the future (as may the development of fuel cells), but strategies for carbon dioxide capture are not deemed economically viable at this time. Within the natural gas industry, officials mentioned using condensers to capture hydrocarbons during the dehydration process as an alternative. Many emitters thought demonstration projects and joint ventures, including involvement by government and educational institutions, would be productive. In Northwestern North Carolina, emitters argued for continued biomass combustion (not a completely technological solution), opposing its imminent curtailment because of pending air-quality regulations.

A surprising consensus exists among the major emitters in awareness of specific strategies and technological solutions for emissions reductions. Less variation is evident between regions than exists between industrial sectors. The locus of control for mitigation is clearly perceived as situated with major emitters, who argued for limiting federal control. Kansans distrust federal initiatives more than do residents of the other three sites, which may be a product of the rural, agricultural nature of the study area or the political culture of the region.

Household perceptions and responses

Household end users, the residents who live and work in the study areas, generate large quantities of greenhouse gases. How do their perceptions of environmental issues correspond to those of the major emitters, and how similar are the household perceptions in the four study areas to national opinions?

How individuals perceive hazards affects the adoption and implementation of effective mitigation programs. Depending on the perception of threat, those at risk will seek additional information, take action to protect themselves, or continue with their normal activities. Hazards not considered threatening, whether due to a lack of information or other factors, result in little or no preparation. Factors that influence the perception of natural hazards include past experience, hazard frequency, cultural and social attributes, and the credibility of hazard information (Burton *et al.* 1993).

Humankind has had a great deal of experience with annual and interannual climate variability and the short-term threats they pose. On the other hand, humanity has had virtually no experience with global warming. If adaptation or mitigation is to take place, individuals must first perceive the issue as a problem or a threat requiring action. One impediment to adaptation or mitigation is the continued challenge to the credibility of hazard information about global warming by skeptical policy makers (Pielke 1994). Human sensitivity to long-term climate change is extremely limited when compared to the perception of short-term

extreme events (Rebetez 1996). Because perceptions are based on two different images –
slow climate change versus short-term weather and climate variation that encompasses
naturally occurring anomalies (Stehr and von Storch 1995) – people react primarily to extreme events, which are often viewed as evidence of long-term climate change. As Rebetez
(1996: 507) notes, 'human perception of climate is strongly influenced by expectations,
which may have little relationship to the true nature of climate as provided by the instru-
mental record.'

National opinion polls since the 1970s have demonstrated a steady increase in public
awareness of environmental issues and, more importantly, support for environmental protection (Dunlap 1992). In the mid-1970s, for example, nearly a third of the United States
population believed that environmental regulations and laws did not adequately protect
the environment. Twenty years later, this percentage had risen to nearly half (Council on
Environmental Quality 1996). As an overall indication of concern, the change is encouraging. In an international poll, Dunlap et al. (1993) found that environmental problems
are salient and important issues in both rich and poor countries with very little difference
between the two in levels of concern. As they conclude, 'personal experience, combined
with increased awareness of the global impact of human activities, has likely made people
around the world begin to recognize that their welfare is inextricably related to that of the
environment' (Dunlap et al. 1993: 38).

Is this pattern of concern replicated for a singular environmental issue, such as global
warming? Despite its prominence in the mass media, relatively few studies on the
public perceptions of climate change are available. A majority of United States citizens
have heard about global warming and generally favor governmental policies to slow or
eliminate its consequences (O'Connor et al. 1999). Other research suggests that such
support may be more symbolic than indicative of willingness to sacrifice individually to
ameliorate such problems (Bord, et al. 1998). One poll suggests that people are aware
of global warming, but do not place it atop their list of environmental concerns, although air pollution, which is thought to drive global warming, is mentioned frequently
(Cushman 1997).

Few studies have examined regional variability in environmental concern, especially
about issues framed as global in scope, such as climate change. As a result, perceptions of
the global warming threat may be related more to misconceptions and perceived ambiguities
about atmospheric science, the nature of scientific uncertainty, and conflicting information
about possible impacts than anything else. The following points summarize the key findings
from the literature (Berk and Schulman 1995; Bord et al. 1997; Bostrom et al. 1995;
Kempton 1991a, b, 1997; Kempton et al. 1995; Read et al. 1994).

- People misconceive global warming and basic atmospheric science. The public equates
  ozone depletion with the greenhouse effect or confuses the greenhouse effect with weather
  and climate or with El Niño.
- Local residents possess variable knowledge about the causes of global warming. Deforestation, industrial emissions, and automobiles are perceived as leading causes, although
  aerosol spray cans, pollution, and ozone depletion are also thought to be important.

- The linkages between cause and effect are poorly understood. The public is aware of possible impacts (e.g. sea-level rise), but does not understand underlying processes; this lack of an identifiable villain creates dissonance, thus diminishing a sense of personal responsibility for greenhouse gas emissions.
- Public evaluations of policy options often are inconsistent with prevailing scientific opinion. For instance, the public believes that reducing deforestation offers an effective strategy for reducing global warming.
- Few realize that human behavioral changes also will be required.

One of the most important conclusions from social science research is that perceptions of the causes of environmental degradation are not linked (conceptually or geographically) to local activities but to external entities such as big business. This lack of culprit creates a dissonance that induces people to downplay their personal roles in causing environmental problems, which in turn makes government intervention more attractive than personal behavior modification. As many pundits have observed, 'We have met the enemy and it is us!' Confounding this disconnection are the misconceptions noted above, although the level of concern about global warming seems to increase regardless of the availability of accurate facts or misinformation (Ungar 1992). Level of knowledge about climate change is a good predictor of an individual's intent to undertake some action in response to it (O'Connor et al. 1999). Willingness to engage in mitigation actions or adaptive strategies is extremely limited, however, and contingent on individual and local contexts (O'Connor et al. 1999).

National media surveys mask local differences in perception and willingness to take action. Disaggregation of public attitudes from the national scale to regional or local venues can help to determine levels of awareness and knowledge, receptivity to action, and constraints on proposed policies. Similarly, not all stakeholders react in like manner, and the opportunities for mitigation may be attenuated or amplified by local conditions and norms. A survey of these local perceptions (major emitters and households) can provide much needed ground truth as regards the likelihood of success for greenhouse gas emission reduction policies and strategies.

Comparisons with national perceptions and responses

Most opinion and social surveys demonstrate that environmental concerns are salient for around a fifth of the issue public, the segment of the electorate that rates environmental concerns as extremely important. When ranked among other issues, environment places third behind abortion and government social services programs, but ahead of gun control and rights for women (Krosnick and Visser 1998). In a 1997–1998 survey of public opinion on global warming, the issue rose from 8% in the September–October survey to 11% in the December–February survey of the public's attention, ranking ahead of race relations and unemployment as salient concerns. As Krosnick and Visser (1998: 4) point out:

This change means that approximately 7.5 million Americans joined the global warming issue public . . . So having the ranks of the most activist segment of Americans on global warming rising by nearly 40 percent is more significant than it might at first appear, because the most vocal and influential segment of the public on this issue nearly doubled in size.

Table 9.1 National and local attitudes toward global change issues

Question	National Agreement	Study Areas Agreement
Scientific consensus regarding climate change exists	36	26
Climate change is a very serious issue	34	30
The effects of climate change will be bad	61	38
Willing to pay more for utilities to prevent climate change	73	29

The Global Change and Local Places sample differs dramatically from those summarized above. Global Change and Local Places households ranked slowing the rate of global warming as their least important social or personal goal (eleven out of eleven). Personal goals (a sound marriage, economic security) rate the highest on a scale of one to seven, with social concerns such as reducing poverty and eliminating the federal deficit at the bottom. Local social goals such as school improvement and lowering the rate of violent crime rank among the top five. Other environmental concerns (air and water pollution, preservation of natural parks) fall midway on the scale in all the study sites, with the exception of Central Pennsylvania, where they were ranked tenth. Little variation was evident among the four study sites in rankings of the most important goals: personal happiness, economic security, lower violent crime, and improving the nation's schools.

Comparing local and national opinions

The Global Change and Local Places surveys included five questions derived from a number of national surveys on climate change perceptions in order to provide comparisons between local and national opinions and perceptions. In most instances, Global Change and Local Places respondents were more skeptical about climate change and its impacts than those who responded to the national surveys. Considerable consensus exists regarding the gravity of the issue and the state of the science, however. National surveys reveal a public aware of substantial disagreement among scientists over whether or not global warming is happening; only 36% of one poll's respondents (Table 9.1) reported scientific consensus on the issue (Krosnick and Visser 1998). Twenty-six percent of Global Change and Local Places respondents believe that energy use causes global warming, while the same proportions judge the relationship to be unproven. One third of Global Change and Local Places respondents believe the truth lies somewhere in between the two positions. Some variability was evident among Global Change and Local Places study areas: more Northwestern North Carolina study area residents take the relationship as fact than do those in other sites (Table 9.2). The Global Change and Local Places sample mirrors national opinion with respect to the existence of global warming.

One third of respondents to national polls acknowledge the gravity of global warming, deeming it a very serious threat (Sustainable Energy Coalition 1996). The Sustainable

Table 9.2 Study area attitudes toward global change issues

Question	Percentage Agreement in Southwestern Kansas	Percentage Agreement in Northwestern North Carolina	Percentage Agreement in Northwestern Ohio
Scientific consensus regarding climate change exists	21	29	27
Climate change is a very serious issue	26	33	30
The effects of climate change will be bad	37	43	35
Willing to pay more for utilities to prevent climate change	30	30	20

Energy Coalition national poll also determined that three quarters of its respondents think that world temperatures have been rising and will continue to rise in the future. More than 60% of Global Change and Local Places respondents believe climate change will be a serious problem in the next 50–100 years if no action is taken, while 30% suggested that the problem is already very or extremely serious.

An examination of more specific levels of information, such as the impacts of global warming on the local level or willingness to pay to reduce emissions, reveals that national opinions differ greatly from those held in the Global Change and Local Places study areas. In response to the question 'Overall, would you say the effects of greenhouse gas emissions for your region would be good, bad, or neither good nor bad,' 61% of the national sample (Krosnick and Visser 1998) respondents thought global warming would have negative impacts, and 15% said it would be beneficial. Global Change and Local Places respondents were more uncertain: 34% did not know (compared with 24% nationally), and 38% thought the effects would be negative. Only 4% foresaw beneficial effects, while the remaining 24% replied neither or no response. Respondents in Northwestern North Carolina were more likely (43%) to foresee negative outcomes than those in Northwestern Ohio and Southwestern Kansas, at 35 and 37%, respectively (Table 9.2).

Willingness of respondents to pay for mitigation is also a facet of opinion on which the Global Change and Local Places respondents differ from their national counterparts (Table 9.1). Nationally, 73% of respondents say they would pay more for utilities to reduce the pollution utility companies produce (Krosnick and Visser 1998). The Global Change and Local Places sample differs sharply, with only 29% of respondents willing to pay an additional US$85.00 per annum to prevent climate change. Willingness to pay also differs among study sites. More Northwestern North Carolina and Southwestern Kansas residents (30%) are willing to incur financial costs to avert climate change than in Northwestern Ohio, where only 20% of respondents would accept such penalties (Table 9.2). Though similar in their understanding of scientific consensus on global warming and its gravity, respondents in our two rural areas – Northwestern North Carolina and Southwestern Kansas – differ in their willingness to pay for prevention programs. Southwestern Kansas residents expressed a greater willingness to pay increased taxes or energy prices in return for emissions reductions, whereas Northwestern North Carolina residents were unwilling to do so. Frequent comments such as 'We pay way too many taxes already' suggest the local reluctance to pay for emission reductions characteristic of Northwestern North Carolina households.

National polls bespeak an environmental consciousness that is not necessarily reflected in localities. Deviations between national and local opinions may result from differences in polling methods, the nature of the sampled populations, the general nature of the questions posed, or combinations of those factors. Global warming is often viewed as a motherhood and apple pie issue, which may lead respondents to wish to appear to pollers to be concerned and responsible. On the other hand, distinctions between national and local opinions may be real in that national averages almost certainly mask local variations in opinion. Clearly, more detailed analyses of subnational opinions should be prerequisite to the formulation of policies and programs to mitigate emissions.

Greenhouse gas emissions: knowledge of effects

Social science literature suggests that the general public misconceives global warming and its impacts, and responses to the Global Change and Local Places surveys support that conclusion. Global Change and Local Places respondents express moderate concern about the general impacts of greenhouse gas emissions, but possess little factual knowledge about impacts or local contributions to global greenhouse gas emissions. Nearly a third of Global Change and Local Places respondents betrayed no knowledge of the effects of greenhouse gas emissions, local contributions to the problem, or local benefits of reduction (Table 9.3). Such a lack of awareness and concern bodes ill for engaging local residents in mitigation programs. As regards policy, whereas a third of the Global Change and Local Places sample offered no opinion concerning the effect of United States government support for international agreements on greenhouse gas reduction, another third viewed such support as affecting their households positively.

Such knowledge gaps reveal a context wherein the local residents fail to link their own behavior to greenhouse gas emissions. This disconnection is even more pronounced with respect to responsibility for greenhouse gas emissions reductions. Because local life is not connected to national or global impacts in the minds of residents, personal responsibility is minimized. Most Global Change and Local Places respondents were equally divided on the question of who was responsible for greenhouse gas reductions (Table 9.3). Though a slim majority thought control rested with individuals, a similar percentage believed the federal government should be responsible for control. Curiously, Northwestern North Carolina respondents also believed more strongly (38%) that the federal government should control emissions, although they also thought (41%) that households were primarily responsible for reducing greenhouse gas emissions.

Table 9.3 Greenhouse gas emissions: scientific knowledge and policy options

Values are percentages for each study site.

Knowledge Item or Policy	Southwestern Kansas	Northwestern North Carolina	Northwestern Ohio	Total
Effects of greenhouse gas emissions on your region of the state				
Positive	5.0	6.4	4.0	5.2
Negative	21.1	34.2	25.9	27.1
Neither	31.2	20.8	31.6	27.7
Don't know/No response	42.7	38.6	38.5	40.0
Compared to the rest of the state, your area's contribution to greenhouse gases is				
More	4.5	9.9	8.1	7.5
Less	42.7	32.7	21.8	32.9
About the same	22.1	28.7	35.1	28.3
Don't know/No response	30.6	28.7	35.1	31.3
Compared to the rest of the state, your area's benefit from greenhouse gas reductions would be				
More	5.5	15.8	8.6	10.1
Less	17.1	14.4	8.1	13.4
About the same	41.7	37.1	50.6	42.8
Don't know/No response	35.7	32.7	32.8	33.7
The United States announced it will support international agreements to reduce greenhouse gas emissions. The effect of this decision on your household is				
Positive	30.7	36.1	33.3	33.4
Negative	11.6	7.9	10.3	9.9
Neither	21.1	22.3	28.2	23.7
Don't know/No response	36.7	33.7	28.2	33.0
Who should be most responsible for the reduction of greenhouse gases?[a]				
Households	29.7	40.6	31.0	33.9
Local government	9.6	11.4	13.8	11.5
State government	7.0	9.9	8.6	8.5
Federal government	30.7	38.1	30.5	33.2

[a] Reflects the percentage of respondents who ranked it with a 1 (most responsible), therefore totals do not equal 100.0

Table 9.4 Energy and transportation efficiency: willingness to undertake vs. already implemented

Mean scores derived from a Likert-like scale with 1 = not willing at all to 5 = very willing for all sites.

Actions	Willingness Mean Score	Percentage Already Implemented[a]	Willingness Mean Score			
			Southwestern Kansas	Northwestern North Carolina	Northwestern Ohio	Central Pennsylvania
Use natural gas instead of electric heat	3.9	38.0	4.1	3.7	4.2	2.7
Purchase new, more energy efficient appliances	3.5	19.0	3.6	3.3	3.6	3.1
Purchase a more efficient water heater	3.6	23.6	3.6	3.6	3.7	3.2
Lower the thermostat on your hot water heater	3.7	21.4	3.7	3.7	3.6	3.5
Install energy-saving light fixtures and switches	3.6	11.5	3.5	3.6	3.7	3.6
Install high-efficiency fluorescent lighting in your home	3.3	11.8	3.4	3.3	3.2	na[b]
Install new or thicker house insulation	3.4	15.8	3.2	3.5	3.3	3.0
Adopt solar energy measures	2.9	2.8	2.7	3.1	3.0	2.4
Use weather stripping and storm windows	4.0	32.7	3.8	4.2	4.0	na
Use window shades, reflective or light-colored curtains, window tinting, and/or plant shade trees near your home for cooling	3.9	21.9	3.9	3.9	4.0	na

Purchase/use more efficient power tools	3.4	4.6	3.2	3.6	3.3	na
Purchase a more fuel-efficient vehicle	3.6	15.2	3.6	3.5	3.7	3.0
Car pool	2.6	3.4	2.6	2.6	2.5	2.1
Drive fewer miles per week	3.0	8.3	2.9	3.0	3.1	2.4
Have your vehicle inspected regularly for efficiency, emissions problems or fuel leaks	3.6	14.4	3.6	3.8	3.3	2.9
Drive below the speed limit	3.2	7.7	3.0	3.3	3.2	2.6
Purchase a natural gas or methane-fueled vehicle	2.4	0.5	2.4	2.3	2.5	na
Participate in mass transit	2.3	0.5	2.2	2.3	2.3	1.7
Work at home (via internet, telephone, or mail)	3.2	3.7	3.0	3.3	3.2	na

[a] Only applies to two of the four sites (Kansas and Ohio).

[b] na = action not included in the local survey

Understanding the link between willingness to mitigate and overt action

Though many respondents recognize global warming as a serious issue, translating that awareness into willingness to mitigate depends greatly on local contexts. Moreover, expressed willingness to mitigate often fails to translate into actual behavior that results in emissions reduction.

The attempt to determine local willingness to act to reduce greenhouse gas emissions was framed within the context of United States energy efficiency. Respondents were asked: 'If the United States were required to become 20% more energy efficient, how willing would you be to implement each action listed below?' The scale ranged from 1 (not willing at all) to 5 (very willing).

Most Global Change and Local Places respondents expressed a willingness (mean score of 3.0 or higher) to undertake the listed actions (Table 9.4), but respondents were largely unwilling to consider (either partly or fully) adoption of solar energy measures or devices, car pooling, use of mass transit, and the purchase of an alternative-fuel vehicle. A minority (around 20%) reported that they had already switched to more energy-efficient natural gas or installed weather-stripping and storm windows. Most respondents (69%) judged their homes to be energy-efficient at the time of the surveys. Most respondents (73%) in all the study areas perceived their current vehicles to be energy-efficient. Some variations in regional willingness to consider switching to natural gas and buying more energy-efficient appliances emerged, with Northwestern North Carolina and Central Pennsylvania respondents less willing to consider those options than Southwestern Kansas or Northwestern Ohio residents (Table 9.4). Both Southwestern Kansas and Central Pennsylvania respondents are less likely to adopt solar energy measures than those living in Northwestern North Carolina or Northwestern Ohio. Central Pennsylvania respondents are less willing to drive fewer miles per week or to car pool than respondents from the other study areas. Many of the differences among study areas are statistically significant. Such variations among localities in willingness of residents to take energy-efficient actions will certainly affect the degree to which localities can implement mitigation or emission reduction strategies.

Though understanding community willingness to act is helpful, the degree to which expressed willingness will translate into changes in behavior can and should always be questioned. Nearly two thirds of Global Change and Local Places survey respondents, for example, report having taken no action to reduce greenhouse gas emissions. Again, variation among sites appears. A larger percentage of Southwestern Kansas respondents (73%) report no action compared to 64% of Northwestern Ohio and 61% of Northwestern North Carolina residents.

Relating concern and willingness to reduce greenhouse gas emissions

As expected, willingness to undertake energy reduction measures correlates with levels of concern about global warming; the higher their level of concern, the more willing respondents are to undertake energy conservation measures. This relationship is statistically significant for three study sites, although the strength of the association varies among them (Table 9.5). Residence in a home perceived to be energy-efficient does not correlate with

Table 9.5 Pearson correlations between willingness to adjust energy use, level of concern, energy efficiency, and electricity cost

*Significant at $p > 0.05$.

Correlation	All sites[a]	Southwestern Kansas	Northwestern North Carolina	Northwestern Ohio
Level of concern	0.36*	0.28*	0.46*	0.43*
Home energy efficiency	0.05	0.14*	−0.09	0.14
Home electric bill	−0.15	−0.27	−0.06	−0.07

[a] Data were not available for the Central Pennsylvania study area.

Table 9.6 Pearson correlations between willingness to adjust transportation use, concern, and number of trucks, vans, and sport utility vehicles per person

*Significant at $p > 0.05$.

Correlation	All sites[a]	Southwestern Kansas	Northwestern North Carolina	Northwestern Ohio
Level of Concern	0.37*	0.26*	0.42*	0.41*
Number of Trucks	−0.10	−0.14	−0.10	−0.06
Number of Vans/SUVs	−0.01	−0.10	−0.22*	−0.01

[a] Data were not available for the Central Pennsylvania study area.

willingness to adjust home energy consumption, except in Southwestern Kansas. On average, residents with smaller electric bills are more willing to undertake energy savings than are residents with higher electric bills. No significant relationship between consumption and willingness to attempt energy savings exists for the other two study sites. Lastly, no significant association is evident in the Global Change and Local Places surveys between willingness to adjust home energy use and dwelling square footage, number of rooms, ownership status, or number of appliances (dryer or air conditioner).

Global Change and Local Places investigators also examined household willingness to adjust transportation behavior and found patterns similar to those encountered in energy use adjustments. For example, strong levels of concern were significantly correlated with respondent willingness to undertake transportation adjustments such as driving fewer miles, buying a fuel-efficient vehicle, and driving below the speed limit (Table 9.6). Little association is evident between transportation adjustments and owning a fuel-efficient vehicle, the number of drivers per household, total miles driven, mean miles driven per household, or the number of cars per household. There is, however, a statistically significant correlation between the number of trucks and sport utility vehicles. Although the association was not strong, it does suggest that higher willingness scores were correlated with fewer trucks, vans, and sport utility vehicles in the household. Conversely, the more trucks and sport utility vehicles in a household, the less likely that household is to undertake transportation-related greenhouse gas reduction measures. Though this association exists for all sites, it is especially pronounced in Northwestern North Carolina.

Regional understandings of household behavior

As has been noted often above, major differences are evident in attitudes about greenhouse gas emissions and reductions among the localities examined by Global Change and Local Places researchers. Not only does variability among localities exist, but even when aggregated, these local opinions often differ from those captured in national polls. The following vignettes about household behavioral patterns in each of the Global Change and Local Places study areas highlight such differences.

Southwestern Kansas

Residents of Southwestern Kansas hold a wide range of opinions regarding climate change (Harrington 2001). Skepticism regarding human impacts on climate and the effectiveness of efforts to ameliorate its effects seem widespread. One resident wrote, 'I don't know a lot about the global warming. Kansas weather does what it wants to do – goes from one extreme to the other – has for years. I doubt that it will ever be predictable. If global warming changes things, people will adjust.' Another resident replied, 'Need more proof of climate change. In the 70s colder than normal, going to have an ice age. Now warmer than normal.' Kansans also distrust government and resent any activity seen to be a waste of tax dollars. Comments such as 'I don't believe there is global warming. I believe this is just a way to try to control people by the government and global warming is not a proven fact – it is part of the liberal political agenda' typify this perspective.

On the other hand, a significant component of the population expresses sincere environmental concerns and a deep sense of personal responsibility. While climate change is somewhat of a worry, it is more nebulous than other environmental issues in the minds of residents; climate change is viewed as less certain and not primarily a local matter. Climate change is thought to be less immediate and a problem that Southwestern Kansas might not have to address, to the degree that it is perceived as primarily an urban phenomenon. Southwestern Kansas residents do, however, recognize the need for action and cooperation.

Northwestern North Carolina

The most striking finding in this study area is an apparent inconsistency between perception and behavior. Almost 40% of Northwestern North Carolina respondents opine that individual households are responsible for greenhouse gas reductions. Though Northwestern North Carolina respondents profess the strongest levels of concern about global warming, they are least likely to undertake such individual adjustments as reductions in home energy use or transportation alternatives in response to their concerns. Most homes are heated by electricity as only the easternmost counties in the study area are served by natural gas pipelines; switching to natural gas was an option for fewer than one third of the respondents to the Global Change and Local Places survey. In general, fuel switching is not a realistic adjustment for this study area. Finally, nearly 45% of the respondents drive a truck, van, or sport utility vehicle as their primary means of transportation, and they are reluctant to alter transportation habits to reduce greenhouse gas emissions.

Northwestern Ohio

Differences between urban and rural residents are evident in the region, with rural residents exhibiting greater skepticism than urban dwellers to the idea that fossil fuel consumption causes global warming. Also, rural residents perceive the problem to be less serious than their urban counterparts and show less concern for global warming as a generational issue affecting their children and grandchildren. Rural residents think that the locus of control should rest with individuals, while urban residents attribute greater responsibility to state and federal governments for greenhouse gas reductions. Northwestern Ohio residents prefer technological approaches to behavior changes, for example, using natural gas instead of electricity for heat, weather-stripping, and purchasing more energy-efficient water heaters. They were less willing to consider car pooling, driving fewer miles, or relocation. The propensity of Northwestern Ohio residents to prefer technological solutions to mitigation and adaptation is a view also held by major industrial emitters in the region.

Central Pennsylvania

The Central Pennsylvania household survey differed in content and timing from those conducted in Northwestern North Carolina, Southwestern Kansas, and Northwestern Ohio. Consequently, the Central Pennsylvania survey results have been treated separately and integrated into the entire sample only when comparable questions were asked. Despite such differences, however, the Central Pennsylvania survey corroborates many findings from the other sites, and it augments those surveys by supplying additional information on how local context influences perceptions of global warming.

Half of the Central Pennsylvania sample was similar in intent to the other surveys in that it addressed local viewpoints on greenhouse gas emissions and mitigation. Unlike the survey used in the other three study areas, however, the second half of the sample framed questions on a national, rather than a local, context. To help emphasize this difference, the cover of the local survey displayed a map of the five-county Central Pennsylvania study region, while the national survey cover depicted a map of the United States. The local survey questions directed respondents to consider local implications of emissions and mitigation, whereas national questions took a more general perspective. For example, one local question asked respondents whether they thought a 3 °F temperature increase would cause starvation and food shortages in Central Pennsylvania. The national question asked respondents if they thought a similar temperature rise would cause starvation and food shortages in the United States. The respective accompanying cover letters also explained the goals of the surveys in corresponding local or national contexts.

To achieve the local–national contrast, it was necessary to change questions slightly from the surveys administered in the Northwestern North Carolina, Southwestern Kansas, and Northwestern Ohio study areas. Though this rewording eliminated the possibility of direct quantitative comparison of the Central Pennylvania results with those from the other regions, indirect, qualitative comparisons can validly be drawn.[1]

- Global warming reduction ranks low as an important goal of respondents when compared to personal interests such as money and marriage. Only 3.5% deemed 'Slowing the rate of global warming' to be their most important goal, while 36.4% ranked 'having a secure and loving marriage' of greatest importance.
- Although 58.3% of the respondents considered themselves moderately to very well in-formed about global warming, they exhibited common misconceptions about the causes of global warming, ranking ozone depletion, rainforest loss, and pollution higher than fossil fuel use, car use, and home heating as main contributors to global warming.
- To reduce greenhouse gas emissions, respondents would prefer to modify their personal behavior rather than support environmental regulation. Though they favored all personal measures for greenhouse gas reduction, at the same time more than 81% thought stronger environmental regulations would hurt them personally, and 70% viewed such regulations as a threat to their jobs.

Beyond these general findings, the survey revealed major differences between levels of concern and willingness to mitigate greenhouse gas emissions elicited by the locally framed and the nationally framed responses. Respondents to the local version of the survey exhibited less concern about the potential effects of global warming in Central Pennsylvania and were less willing to modify personal behavior than those responding to the national versions of the same questions. Moreover, respondents to local questions were less worried about the potential effects of global warming regardless of where those effects would be felt – in Central Pennsylvania, the United States, or the world.

The results suggest that individuals perceive global warming and therefore greenhouse gas emissions as a global problem that is best addressed at the national level. Although there is concern for the problem and recognition that individuals should be responsible for helping solve it, respondents do not perceive the impacts of the human activities that produce greenhouse gases to be local issues. Abundant research demonstrates that global warming's impacts will be local and felt locally, whereas Global Change and Local Places research documents the need to realize reductions in greenhouse gas emissions through

---

[1] The slightly different agenda of the Central Pennsylvania research team resulted in survey implementation that varied from the procedure used at the other three study sites. The sample – acquired by subcontract – consisted of 1,200 randomly chosen residents of the five-country study area. In early December 1998, each of these residents received the first mailing. This mailing contained a professionally printed survey booklet, a cover letter on university stationery, an addressed and stamped return envelope, and a dollar bill as incentive to complete the survey. Six hundred packets contained the local survey and 600 contained the national instrument. Reminder postcards were sent in mid-December. In mid-January 1999, a second packet essentially the same as the first was mailed, but did not include the dollar incentive. By March 1999, 647 surveys were returned for analysis, a response rate of 62.75% after accounting for bad addresses.

action at the locality scale. Resolving the disconnection and inconsistencies between global perceptions and local realities appears to be a daunting task.

Prospects for emission reductions

Recognition of the existence of human-induced climate change is variable among the Global Change and Local Places households surveyed in the Global Change and Local Places study areas. When climate change is recognized as an issue, its salience among other concerns remains low. These locality findings stand in sharp contrast to national opinion polls that document higher levels of concern and greater willingness to reduce emissions (even at costs to individual households) than was evident in the Global Change and Local Places study areas. Key actors and households hold mixed opinions regarding the impacts global warming could have on their localities. Most major emitters in Northwestern Ohio, for example, believe that anthropogenic causes for global warming remain questionable, and that mitigation policies should accordingly be applied cautiously. A need to act immediately is not evident to most key actors and to the majority of households and stakeholders in Northwestern Ohio, who perceive the problem to have import largely for future generations. In the Southwestern Kansas and Northwestern North Carolina study areas, households and major emitters demand proof that a problem exists as a prerequisite to action that would reduce emissions.

Both emitters and households prefer technological fixes to strategies for reducing emissions that require changes in behavior. That preference should occasion no surprise, given the limited understanding of the connection between households and emissions and between emitters and warming at the global scale. In Northwestern Ohio, the preference for technological solutions was evident among both households and major emitters. Other sites revealed more mixed opinions, although preferences still inclined generally toward technological solutions such as fuel switching. Rarely did emitters or households volunteer to change their behavior by driving less or by using solar energy. On the contrary, major emitters have little faith in behaviorally based strategies for curbing consumer demand for energy. The increased popularity of such fuel-inefficient vehicles as trucks and sport utility vehicles in the Global Change and Local Places study areas, for example, bodes ill for behavioral strategies for reducing greenhouse gas emissions.

Respondents in both categories prefer local action with federal oversight to top-down federal mandates. Major emitters are inclined to resist federal attempts to reduce or control greenhouse gas emissions, though some consensus exists that federal agencies can be effective in setting broad goals and in providing coordination and oversight. Respondents have greater faith in local institutions and officials than in federal agencies and managers, resulting in the view that the locality offers the most appropriate setting for seeking emissions reductions. Paradoxically, respondents also doubt the ability to find the expertise necessary to effect greenhouse gas emission reductions in their localities, and also are somewhat skeptical that local impediments to emissions reductions rooted in self-interest can be overcome.

Many commonalities are evident among local attitudes toward greenhouse gas emissions and emission reductions expressed by major emitters and households included in the

Global Change and Local Places surveys in the four study areas. The differences between attitudes captured in the Global Change and Local Places localities and those revealed by national surveys are somewhat larger than the Global Change and Local Places research team expected, as are the variations in attitudes among the study areas. Greenhouse gas emission reduction policies and strategies will ultimately succeed or fail at the locality scale, where the decisions that will effect or fail to effect reductions must be made. Policies and strategies will be effective largely to the degree that they can be tailored to local political and social contexts. Top-down federal mandates for reductions have little likelihood of altering emitter or household behavior in Global Change and Local Places localities or elsewhere.

REFERENCES

Berk, R. A., and D. Schulman. 1995. Public Perceptions of Global Warming. *Climatic Change*, **29**: 1–33.

Bord, R. J., A. Fisher, and R. E. O'Connor. 1997. Understanding Global Warming and Promoting Willingness to Sacrifice. *Risk*, **8** (4): 339–54.

Bord, R. J., A. Fisher, and R. E. O'Connor. 1998. Risk Perceptions, General Environmental Beliefs, and Willingness to Address Climate Change. *Risk Analysis*, **19** (3): 461–71.

Bostrom, A. M., G. Morgan, B. Fischhoff, and D. Read. 1995. What Do People Know about Global Climate Change 1. Mental Models. *Risk Analysis*, **16** (6): 959–70.

Burton, I., R. Kates, and G. F. White. 1993. *The Environment as Hazard*, 2nd edition. New York: Guildford.

Council on Environmental Quality. 1996. *Environmental Quality: 25th Anniversary Report*. Washington, D.C.: United States Government Printing Office.

Cushman, J. H. Jr. 1997. Global Warming Fight Wins Wide Support. *New York Times*, November 28, A14.

Dunlap, R. E. 1992. Trends in Public Opinion Toward Environmental Issues: 1965–1990. In R. Dunlap and A. Mertig, eds. *American Environmentalism*: 89–116. Philadelphia: Taylor & Francis.

Dunlap, R. E., G. H. Gallup, Jr., and A. M. Gallup. 1993. Of Global Concern: Results of the Health of the Planet Survey. *Environment* **35** (9): 7–15, 33–9.

Harrington, L. M. B. 2001. Attitudes toward Climate Change: Major Emitters in Southwestern Kansas. *Climate Research*, **16** (2): 113–22.

Harrington, L. M. B., and M. Lu. 2002. Beef feedlots in Southwestern Kansas: Local change, perceptions, and the global change context. *Global Environmental Change*, **12** (4): 273–82.

Kempton, W. 1991a. Lay Perspectives on Global Climate Change. *Global Environmental Change Human and Policy Dimensions*, **1** (3): 183–208.

Kempton, W. 1991b. Public Understanding of Global Warming. *Society and Natural Resources*, **4**: 331–45.

Kempton, W. 1997. How the Public Views Climate Change. *Environment*, **39** (9): 12–21.

Kempton, W., J. S. Boster, and J. A. Hartley. 1995. *Environmental Values in American Culture*. Cambridge, MA: The MIT Press.

Krosnick, J. A., and P. S. Visser. 1998. The Impact of the Fall 1997 Debate about Global Warming on American Public Opinion. *Weathervane*, Resources for the Future's digital forum on global climate policy. http//www.weathervane.rff.org/

O'Connor, R .E., R. J. Bord, and A. Fisher. 1999. Risk Perceptions, General Environmental Beliefs, and Willingness to Address Climate Change. *Risk Analysis*, **19** (3): 461–71.

Pielke, R. A. Jr. 1994. Scientific Information and Global Change Policymaking. *Climatic Change*, **28**: 315–19.

Rayner, S., and E. L. Malone, eds. 1998. *Human Choice and Climate Change*. Columbus: Batelle Press.

Read, D., A. Bostrom, M. G. Morgan, B. Fischhoff, and T. Smuts. 1994. What Do People Know about Global Change. 2. Survey Studies of Educated Laypeople. *Risk Analysis*, **14** (6): 971–82.

Rebetez, M. 1996. Public Expectation As An Element of Human Perception of Climate Change. *Climatic Change*, **32**: 495–509.

Redclift, M. 1998. Dances with Wolves? Interdisciplinary Research on the Global Environment. *Global Environmental Change: Human and Policy Dimensions*, **8** (3): 177–82.

Stehr, N., and H. von Storch. 1995. The Social Construct of Climate and Climate Change. *Climate Research*, **5**: 99–105.

Sustainable Energy Coalition. 1996. America Speaks out on Energy: a Survey of 1996 Post-election Views. http://www.citizen.org/cmep/renewables/11-96pollresults.html

Ungar, S. 1992. The Rise and (Relative) Decline of Global Warming as a Social Problem. *The Sociological Quarterly*, **33**: 483–501.

10

Reducing greenhouse gas emissions: learning from local analogs

C. Gregory Knight, Susan L. Cutter, Jennifer DeHart, Andrea S. Denny, David G. Howard, Sylvia-Linda Kaktins, David E. Kromm, Stephen E. White, and Brent Yarnal

Global change is rooted in localities. The impacts of global warming, as well as adaptations to warming and attempts to ameliorate it, will occur in communities at local and regional scales. In some respects global climate change is analogous to other societal dilemmas. Local communities contribute to large, intractable problems; local initiatives may arise in the absence of larger efforts to address the problems; and localities grapple with policies and regulations imposed upon them from afar. Within each Global Change and Local Places study area, there are human–environment analogs that yield insights into how greenhouse gas mitigation could proceed at locality scale.

Earlier chapters have documented the import of understanding the driving forces that generate regional greenhouse gas emissions, tracked changes in emissions through time, and assessed greenhouse gas abatement potentials in the Global Change and Local Places study areas. Examining the structures and dynamics of societal attempts to mitigate threats analogous to global warming offers fresh insights for science and policy formulation. Whereas prior work has focused on using analogs to anticipate impacts and adaptation,[1] this chapter emphasizes analogy to understand mitigation processes. For purposes of this analysis, *adaptation* denotes the array of societal coping responses to an environmental threat such as climate change. *Mitigation* means efforts to abate the threat itself, such as limiting the release of greenhouse gases or acting to absorb them.

Local analogs for greenhouse gas mitigation

Human–environment analogies exist in each of the Global Change and Local Places study areas that offer insights into how reduction of greenhouse gas emissions could be

---

[1] Using analogs to study climate change impacts was proposed by the Working Group on Impacts at the World Climate Conference in 1979 (Glantz and Ausubel 1988). Analogies from past climatic events have been used to analyze current situations. Easterling *et al.* (1992) used the 1930s drought in the Great Plains to assess possible impacts of such a climate event on contemporary agriculture and the regional economy of four states in the American Midwest. Analogs have also been used to learn how society responded to past climatic or climate-related environmental threats. Examples include adaptation to sea and lake level rise, drought, natural hazards, urban climate, the collapse of ancient civilizations, the green revolution, and population growth (Glantz 1988, 1991; Meyer *et al.* 1998; Kates 2000). The principal use of analogs in greenhouse gas mitigation involves comparisons with ozone depletion or acid rain abatement (Allen and Christensen 1990: Kowalok 1993; Solomon 1995: Kerr 1997, 1999; Clark *et al.* 2001).

Table 10.1 Elements of local environmental analogs to greenhouse gas mitigation

1. Local actions contribute to an environmental problem over which the locality has little control
2. Evidence of early local concern or initiative exists; may precede higher level and wider commitments
3. Local mitigation alone has little impact on the major problem
4. Local mitigation may offer some local benefit
5. Local adaptation to impacts of the problem can occur
6. Higher-level and wider commitments appear in voluntary form
7. Higher-level and wider commitments appear in regulatory form
8. Local response and initiative addresses wider voluntary and regulatory imperatives
9. Collective and local benefit from addressing the problem emerges
10. As the problem moves toward resolution, there are winners and losers

achieved locally (Table 10.1). First, a locality contributes to a widespread problem such as greenhouse-gas-induced climate change. As in the case of communities that have joined the Cities for Climate Protection initiative (ICLEI 2002), local concern and initiative arise, but these local initiatives alone cannot address the larger problem, even though local efforts may produce some local benefits. (In the case of greenhouse gas mitigation, for example, there may be some 'no-regret' benefits from energy conservation, but isolated local efforts are insufficient to mitigate global warming.) Voluntary commitment begins to appear at wider scales, for example in nations committing to the Kyoto Accords and in industrial organizations, followed by such regulatory actions as emissions trading or carbon taxes in the greenhouse gas emissions case. Local initiatives take place within wider voluntary and regulatory imperatives and eventually the community realizes benefits from collective action, such as avoiding adverse potential consequences. Yet, in the process and solution there are winners and losers, individuals and entities who suffer real or potential costs that are not paid by those who gain.

The four analogs from the Global Change and Local Places sites are:

- depletion of the Ogallala Aquifer from the Southwestern Kansas study area;
- watershed protection from the Northwestern North Carolina study area;
- Lake Erie water pollution from the Northwestern Ohio study area; and
- the decline of the regional coal economy from the Central Pennsylvania study area.

The Southwestern Kansas and Northwestern Ohio cases illustrate local responses to broader environmental imperatives. The Northwestern North Carolina case provides insight on the use of state- and local-level government mandates to address a widespread occurrence of a common local environmental problem. The Central Pennsylvania case suggests that the negative impacts of acid rain mitigation on regions producing high-sulfur coal could be directly compared to the consequences of decreased coal production resulting from greenhouse gas mitigation. These Global Change and Local Places analogs play out in different ways, but they run sufficiently parallel to yield valuable insights into potentials for local action to reduce greenhouse gas emissions. Taken together, the four examples illustrate how mitigation comes about, what local costs and benefits arise in the process, and what dimensions of change occur when costs are transmitted from other regions.

Depletion of the Ogallala aquifer

Exploitation of Ogallala water is the major force driving landscape change in Southwestern Kansas. Several analysts have examined the Ogallala Aquifer in connection with human adaptation to climate change (Warrick and Riebsame 1983; Glantz and Ausubel 1988; Wilhite 1988; Glantz 1992), viewing its depletion as a useful example of societal resilience in the face of environmental, societal, and technological changes. Like global warming, groundwater depletion is a gradual, continuous, long-term, cumulative process impossible to reverse over the short term. Can and will local residents mitigate human-induced environmental changes with clear adverse consequences? Institutional and local initiatives to slow rates of groundwater depletion suggest the answer is yes, and the ways local residents have responded to aquifer depletion point the way to likely adjustments and responses to global warming.

Irrigation and groundwater depletion

Like much of the High Plains in which it is situated, the six-county study area consists largely of treeless grassland with relatively modest relief and limited rainfall (Webb 1931). The study area lies within the center of the 1930s Dust Bowl. Water is often scarce, and much of its agriculture is now dependent upon Ogallala water. Irrigation using water from the Arkansas River began as early as the 1880s. Through World War II, most irrigation water in the region was pumped from the rivers or artificial lakes into fields via narrow canals or ditches. Since 1950, irrigated agriculture and regional economic viability have become increasingly based on water pumped from the Ogallala aquifer. Aquifer depletion has been a major concern in the area for at least 25 years.

When high-capacity pump engines first made lifting underground water possible on the High Plains, it was transported onto level fields by pipes that emptied the water into furrows between crop rows, whence it flowed across the fields. Center pivot sprinkler systems were introduced in the late 1960s. These wheeled units that circle a central point can traverse uneven and hilly land. By 1978, center pivots were applying water to an estimated 27% of High Plains irrigated land. Center pivot sprinklers need less labor than pipe systems, and they distribute water more uniformly (Kromm and White 1981). Water use accelerated as the adoption of center pivots greatly expanded the amount of land that could be irrigated. Irrigated area peaked in the late 1970s and then declined through the 1980s, to be followed by a rebound in the 1990s that still continues. Irrigated area in the study area expanded from 7,500 hectares (18,500 acres) at mid-century to 375,000 hectares (926,000 acres) in the mid-1990s.

Cropping patterns changed in response to more abundant water. Corn and wheat were sown at the expense of other crops. By the late 1970s, High Plains residents created a new flourishing Western Corn Belt based on underground rain (Green 1973). Throughout most of the region, farmers raised forage crops to feed cattle. The cattle industry centered on huge feedlots that collectively consumed many tons of grain daily. The cattle in turn were sent to meat processing plants, some of which were among the largest in the world. Irrigated agriculture undergirded almost all aspects of the local agribusiness economy. Where irrigation prevailed, rural towns were vital and healthy. Ogallala water appeared to have brought stability to a large part of the High Plains, including Southwestern Kansas.

Or had it? Depletion of the aquifer has been a nagging concern in Southwestern Kansas throughout the past 25 years of prosperity. Ogallala water is fossil water trapped in an aquifer that underlies 440,000 square kilometers (170,000 square miles) of the High Plains. The Ogallala gets minimal recharge, and those pumping it are essentially mining groundwater. Recharge varies locally, but the long-term average rate for the entire High Plains is less than a centimeter per year (Gutentag *et al*. 1984). For all intents and purposes, over most of the High Plains (including the study area), water pumped out of the Ogallala is gone forever. Moreover, depth to water varies significantly throughout the High Plains and over smaller areas such as counties. An irrigator pumping water from 30 m (98 ft) spends about a third of the energy needed by an irrigator who must pump from a depth of 90 m (295 ft) (Gutentag *et al*. 1984). Irrigators throughout the High Plains are often more concerned about the energy costs of pumping water than about the threat of depletion (Kromm and White 1985; Taylor *et al*. 1988).

National and regional responses

Responses to groundwater depletion in the High Plains range from individual actions to federal government projects. Although responses now seem comprehensive, a long period of ignorance or denial preceded adjustment and mitigation. As late as the 1970s, many High Plains residents were unaware of groundwater depletion (Knight and Rickard 1971), and some believed the Ogallala aquifer to be a large underground river or lake fed by Rocky Mountain snowmelt. In response to growing concern over groundwater depletion, the United States Congress in 1976 authorized a Six State High Plains – Ogallala Aquifer Regional Resources Study of Texas, New Mexico, Oklahoma, Colorado, Kansas, and Nebraska. Its intent was to 'examine the feasibility of various alternatives to provide adequate water supplies for the High Plains region' and to 'assure the continued economic growth and vitality of the region' (High Plains Associates, Inc.1982: 1). Completed in 1982 at a cost of US$6 million, the study concluded that irrigated agriculture and the agri-economy in the six states would expand from 5,700,000 hectares (14,079,000 acres) in 1977 to some 7,500,000 million hectares (18,525,000 acres) in 2020. This expansion would be offset, however, by a decline of 227,000,000 cubic meters (18,900 acre feet) of water (Kromm and White 1986).

Many experts consider the study to be a benchmark assessment of irrigated agriculture in the American High Plains. Although it was a massive undertaking, its limitations cannot be ignored. The study's aggregate, macro-level approach conceals local differences in groundwater policy, management, conservation, and irrigation practices (Kromm and White 1986). The High Plains aquifer cannot be managed as a single entity. Six states with widely varying water laws and physical conditions are involved, and no uniformity in depth to water, saturated thickness, soil type, or length of frost-free season exists among them. Imposing federal policies in this politically and physically fragmented region is unlikely to succeed. State and local controls conform more closely to regional conditions and preferences (Kromm and White 1986).

State and local responses

Although Kansas surface waters were brought under the prior appropriation doctrine in 1886, groundwater was not placed under state control and subjected to prior appropriation until the Kansas Water Appropriation Act of 1945. The act decreed 'first in time, first in right' for groundwater, but also recognized existing or vested water rights (Smith 1983). All water rights for beneficial use were to be secured through a permit system that established the priority of rights. Local control through groundwater management districts began with the Kansas Groundwater Management District Act of 1972, which enabled qualified local voters to form a district in consultation with the chief engineer of the Kansas Division of Water Resources. Five districts have been established, three of which manage the High Plains aquifer in western Kansas. Each district has extensive authority including eminent domain and the authority to levy water charges, require water metering, govern well spacing, and recommend rejection of requests for new wells (White and Kromm 1996). Kansas groundwater management districts exercise greater authority and have more autonomy than those in other states of the High Plains (White and Kromm 1995).

In 1992, a joint state and local effort officially known as the Kansas Agriculture Ogallala Task Force was created by the Kansas State Board of Agriculture. Most of its 23 members were regional irrigators, ranchers, and community leaders. The Task Force was charged to 'investigate ways of maintaining and enhancing the agricultural economy of the Ogallala region as it impacts the state of Kansas while lessening irrigated crop demands on the aquifer' (Kansas State Board of Agriculture 1993: 1). In addition to creating a high level of public awareness and engaging many people in its monthly meetings, the Task Force generated renewed interest in research to improve irrigation efficiency. In 1993, the task force issued an extensive set of recommendations, most of which focused on extending the life of the Ogallala aquifer.

In 1979, Kromm and White initiated a series of studies on responses to groundwater depletion in the High Plains Ogallala region (1981, 1983, 1984, 1990, 1991, 1992; White and Kromm 1995). Several large sample surveys elicited farmer and public views on the saliency of groundwater decline, preferences for specific adjustments, mitigation strategies, water conservation techniques employed, and opinions regarding the levels of government that should take responsibility. In addition, numerous field interviews were conducted with farmers and knowledgeable professionals. All six counties in the Global Change and Local Places study site were included in two of the surveys, and Finney, the largest and most populous county, was included in the other two.

Strategies that allow extensive use of Ogallala water but compensate for it include the importation of water from other states, adoption of weather modification to increase rainfall, construction of small recharge dams, air injection to increase pumpable groundwater, and maintaining minimum base flows in perennial streams. Most of these strategies require external funding, and respondents expressed concerns about the cost of technological fixes, fears about who would pay, and distaste for dependence on federal and state governments for cost sharing, construction, regulation, or control. Within the larger Ogallala region, the preference for strategies that increased the available water varied positively with the degree of groundwater depletion.

In 1980, survey respondents in 13 Southwestern Kansas counties favored (70%) recharge dams (though limited precipitation makes them impracticable), were evenly split (49% favored) on weather modification, and generally disliked (27% favored) importing water from other states (Kromm and White 1981). A 1984 survey of irrigators in three western Kansas counties yielded similar results: 76% favored recharge dams, 51% favored weather modification, and 36% supported water imports.

Overall, the 1980 study identified 31 possible adjustments to groundwater depletion. More efficient water use by irrigators accounted for five of the six most preferred adjustments. There was also strong support for greater equity in water allocation in the form of high acceptance of equitable apportionment and water quotas. The least popular adjustments were financial incentives or disincentives, limiting irrigation, and regulating water use. Nearly every form of government intervention was unpopular. Maintaining the status quo, however, was equally distasteful. The results bespeak a willingness to change, and the preferences expressed held constant across local differences in type of farming and off-farm activity (Kromm and White 1981, 1984).

Additional surveys were conducted in the mid-1980s to assess responses to groundwater depletion over the entire area underlain by the Ogallala aquifer. Again, a very strong preference for irrigator water conservation practices dominated responses: eight of the most preferred adjustments were in that category. Building reservoirs to store water and encouraging water conservation rounded out the most preferred list (Kromm and White 1985). Adjustments least favored were similar to those unpopular in the Southwestern Kansas study: financial incentives and penalties, and management policies that restricted water use were strongly disliked, as was government intervention. Paradoxically, water conservation laws and building reservoirs to store water were seen as proper responsibilities of government.

The 1988 study of High Plains water conservation practices suggests that irrigators do and will adapt. They have already taken measures to reduce rates of groundwater decline in response to decreased supply. In addition, many irrigators understand dryland farming techniques. Although they hope to extend the aquifer's life, many are quick to note that their grandfathers farmed western Kansas without irrigation and they will also, if need be. Indeed, the Ogallala region hosts many more acres of dryland than of irrigated agriculture; only a third of study area acreage is irrigated. Unlike global warming, groundwater depletions suggest a return to a situation experienced by the inhabitants' recent ancestors.

Table 10.2 Common factors in the Ogallala aquifer – greenhouse gas analog

---

Gradual, continuous, long-term, cumulative processes becoming critical and difficult to reverse
Human-induced environmental hazards accumulate with time
Local causes have much wider-scale consequences
Challenges to the long-term sustainability of agriculture in the region
Reduced water availability for agricultural use in the High Plains
Require societal response and adjustment to environmental change
Promote economic uncertainty
Negative impacts on public well-being
Demand mitigation that is effective over a long period of time
Require adaptations of significant consequences if mitigation fails
Depend on the rise and decline of energy prices
Impacts can be mitigated with technological fixes
People opt for short-term gains with long-term costs
Require widespread actions to make most responses effective
Have reciprocal relationships with agriculture

Ogallala depletion and global warming mitigation

Although the Ogallala aquifer is regional and warming is global, the movement of the meat packing industry into the Southwestern Kansas study area and its demands for groundwater have added a global overlay to groundwater depletion in the region. Drawing analogies between groundwater depletion and global warming is appropriate; both are gradual, continuous, long-term, cumulative processes that are increasingly difficult to reverse; both challenge the long-term sustainability of agriculture in Southwestern Kansas. The result is that both promote economic uncertainty and require societal responses and adjustments (Table 10.2).

Irrigators have adopted water-saving practices and their groundwater use has declined. The rate of change in mean depth to water for measured wells in Finney County has declined from −0.44 m (−17.30 in) per annum from 1978 to 1987 to only −0.07 m (2.76 in) per annum between 1992 and 1995 (Woods *et al.* 1995). What has led irrigators to undertake these measures? While assigning high priority to groundwater conservation, irrigators are more concerned about low crop prices and energy costs. In other words, mitigation measures will be undertaken if they yield economic benefits, but neither conservation nor environmental concerns *per se* constitute primary motivations for change.

The degree to which Southwestern Kansas irrigators willingly implement water-saving practices varies with the locations and sizes of their operations. Large operators (five or more irrigation wells) tend to support irrigation efficiency, oppose government subsidies, and resist restrictions on water use. Smaller operators are more receptive to subsidies and restrictions. In contrast to ambiguity about global warming, irrigators are confident groundwater depletion is real. That was not true 30 or 40 years ago, when many irrigators believed the Ogallala to be inexhaustible. Water-saving practices were widely adopted only after 1975, when declining well flows and deeper wells became impossible to ignore. This history suggests that, once irrigators accept global warming as reality, they will adopt mitigation strategies that work to their economic advantage, much as they have done in response to groundwater depletion. Now, however, willingness to mitigate global warming at the local level is weak. Lack of awareness of global warming threats, combined with negative attitudes toward governmental controls, suggest that adaptation rather than greenhouse gas emission reductions will predominate.

There is little agreement on current and the future adaptability of agriculture to a variable and changing climate (Chiotti and Johnston 1995). Easterling (1996) suggests that climate change should not pose insurmountable challenges to North American agriculture. Agronomic adaptation strategies include changes in crop varieties and species, the timing of operations, and land management, including irrigation. Economic strategies include investment in new technologies, infrastructure and labor, and shifts in international trade. Such reactions could '. . . offset either partially or completely the loss of productivity caused by climate change' (Easterling 1996: 1). Past local responses in the Ogallala region suggest that Easterling is correct. The most serious impediment to adjustment is skepticism among Ogallala residents that global warming is occurring. Once credible evidence is in hand, the users and managers of the Ogallala should be willing to adjust their practices to deal with the impacts of global warming on their operations.

North Carolina watershed protection

Water quality issues have a long and complex history within the United States and North Carolina (Howells 1990; Stoloff 1991). Early concerns focused on public health, with numerous governmental programs that sought to control water quality through identification of point pollution sources such as sewage and industrial wastewater discharges, the establishment of effluent limits, and applications of technology to clean water supplies or to protect them from pollution. To safeguard drinking water supplies during the nineteenth century, entire watersheds were acquired by municipalities and preserved from human activity. In the study area, that action was taken by the City of Asheville (Howells 1989, 1990; Moreau et al. 1992). The exclusion approach to water quality was distinctive in that it encompassed not only water, but also the entire surface area that contributed to water supplies.

For a time, advances in water treatment and urban and industrial growth led to less emphasis on controlling watershed land use as the primary means of protecting water quality. Filtration and chlorination came to dominate water treatment. But as urbanization and industrialization accelerated, toxic chemicals in the environment multiplied and water treatment technology was unable to keep pace with pollutants. A 1969 survey by the United States Public Health Service Bureau of Water Hygiene found just over half of United States community water supply systems met then current drinking water quality standards (Moreau et al. 1992). As a result, planners began to look once again to watershed programs as ways to assure safe water. Initial watershed protection strategies sought to limit or prohibit human activities in water source areas. As cities expanded, however, excluding all human activities from entire watersheds became less and less practicable.

Today's watershed protection programs try to control types and density of land use, storm water flows, erosion and sedimentation, agriculture, forestry, and wastewater discharge rather than proscribing such activities within drinking water catchments. Coupled with pretreatment of industrial wastewater, emergency responses to accidental spills, watershed sanitary surveys, water quality monitoring, and assurance of minimum stream flows, the watershed protection approach has been widely adopted (Howells 1990: 102; Dilks and Freedman 1994; Brady 1996; Davenport *et al.* 1996). By recognizing the nested nature of watersheds, specific actions can be taken at state, basin (regional), or watershed (local) scale as needed. Some water quality issues (phosphate loading) can be addressed only across entire basins. Other problems such as permitting are best resolved by statewide measures. Quality monitoring or controlling nutrient loadings of individual water bodies is best pursued locally.

The watershed protection approach goes beyond water quality and human health concerns to focus on biological integrity and habitat issues. Because these concerns are so diverse, multiple agencies must cooperate to achieve successful watershed protection (United States Environmental Protection Agency 1995). Establishing watershed protection requires collaborative setting of priorities, marshaling stakeholders and managing agencies, and constant monitoring to assess the success of adopted policies. The holistic nature of the watershed protection programs makes them an appropriate analog for greenhouse gas mitigation programs.

Events leading to the Water Supply Watershed Protection Act

Legislation to protect watersheds was introduced in the North Carolina General Assembly in 1897, when the State Board of Health asked the legislature to extend the police powers of cities and towns to their watersheds and to require periodic watershed inspections. The bill never reached the assembly floor, however (Howells 1989). More than a century later, the legislature passed the landmark 1989 Water Supply Watershed Protection Act (WSWPA). As late as 1980, a legislative study had concluded that water problems in North Carolina did not warrant bringing the issue to the attention of the General Assembly. Two years later, however, commentators began to link water quality to continued economic growth (Alston 1982; Stewart 1982). At the same time, environmental awareness and activism began to emerge in North Carolina (Oliver 1983). Growing public interest in environmental topics was rooted in 'first hand destruction of their natural resources' (Mather 1988). Environmental groups ranging from grassroots organizations to the Sierra Club experienced dramatic increases in statewide membership during the early 1980s (Effron 1988). North Carolinians mirrored rising national concerns over water pollution, hazardous waste facility location, air pollution, acid precipitation, and the environmental consequences of development (Anonymous 1988; Mather 1988). A 1987 statewide ban on phosphate detergents followed pitched battles between citizens' groups and detergent lobbyists, but by the late 1980s, North Carolinians favored environmental protection even if regulation resulted in increased taxes or slowed economic growth. A statewide poll revealed that almost three quarters of those surveyed opposed economic growth detrimental to the environment, with highland residents most concerned about environmental degradation (German and Hoffman 1992).

Increasing concern about water quality and quantity in the eastern part of the state served as the catalyst for what ultimately developed into the Water Supply Watershed Protection Act (Gray 1992). A search for additional water sources for the Chapel Hill – Carrboro – Orange County area (home of the University of North Carolina) led to the selection of Cane Creek, a relatively undeveloped watershed. When watershed property owners opposed Chapel Hill's plans, a protracted, statewide discussion ensued. The City of Raleigh proposed to control pollution within the extensive Falls of the Neuse Reservoir, upon which it relied, by imposing sweeping land-use controls. The resulting controversy resulted in numerous front-page newspaper stories in the mid-1980s. At the time, pollution control measures were voluntary, and the rapid development of the Triangle area, exemplified by a 2,150 hectare (5,300 acre) residential and industrial project in the Falls of the Neuse watershed, resulted in a 1987 general assembly bill to mandate local government controls in water supply watersheds. The bill failed. It had little support from local governments except in the Piedmont, which relied on surface water and had already encountered the problems resulting from dense population, development, and the resulting pollution. North Carolina's coastal areas rely primarily on groundwater and the state's mountainous western area is sparsely settled, resulting in less experience with pollution. Upland area local governments were also skeptical of the propriety of land-use controls generally (Moreau *et al.* 1988; Gray 1992: 34).

Though the 1987 effort to mandate land-use controls failed, the general assembly did establish a Watershed Study Commission, which it directed to report back to it in 1989 (Gray 1992: 32–33). The commission incorporated opinions and recommendations from a broad array of officials and organizations in the final report to the assembly. The resulting 1989 Water Supply Watershed Protection Act enjoyed strong support in both houses of the legislature and passed easily. With its enactment, statewide minimum standards to protect water supply watersheds were established for implementation in 1991, and all surface water supplies were to be reclassified by 1 January 1992. From the twelve public hearings and period for written comments that ensued, several major issues emerged: the sizes of the areas in which development was to be restricted; whether development density restrictions constituted a taking of property rights; the effects of restrictions on economic development; potential duplication of existing regulations, diminution of local autonomy, and the costs to local government of enforcement. The validity of the WSWPA was questioned in 1996, when the North Carolina Court of Appeals declared it unconstitutional (North Carolina Court of Appeals 1996; North Carolina Department of Environmental and Natural Resources 1999), but the North Carolina Supreme Court overturned the lower court ruling in July 1997 (Ducker 1999).

Seven years elapsed between the first statewide expressions of concern over water supply quality and the passage of the WSWPA in 1989, and another five years passed before full implementation of the act's restrictions was achieved in 1994. Six years later, monitoring systems had yet to be established, which means that any improvements in water quality that may have been achieved cannot be measured.

Table 10.3 Comparison between North Carolina water quality issues and global warming

---

Like greenhouse gases, water pollution takes place locally, but in the appropriate concentrations, it influences areas beyond local watersheds

Both water pollution and greenhouse gas emissions are part of broader processes or systems (natural, social, economic)

Solutions to water quality problems, like global warming and reductions of greenhouse gas emissions, can only be achieved through multi-scaled, comprehensive, and flexible actions

Fear of change is a component in both water quality issues and global warming

Technological fixes are sought in both issues as a more acceptable alternatives to more comprehensive actions, and especially behavior change

Like efforts to address global warming and reduction of greenhouse gases, water quality standards originate at the federal level, but actions to achieve goals are formulated at state levels. Ultimately, regulations affect individuals and organizations at local scales

In global warming, those responsible for the largest greenhouse gas emissions do not necessarily face the greatest impacts of global warming; the same can be true in water quality issues. Those downstream from major polluters frequently suffer the most serious impacts

In water quality issues and global warming there are winners and losers. Ways to compensate those most affected must be devised

Important at the local, regional, and state levels, current water quality issues do not extend to the global scale

---

North Carolina water quality and global warming

If the people of North Carolina come to understand the nature of global warming, their earlier actions on water pollution bode well for efforts to ameliorate production of greenhouse gases. Once aware of water pollution, North Carolina passed the WSWPA. The process leading to watershed protection regulations offers some instructive parallels with global warming (Table 10.3).

The clean-up of Lake Erie

Erie is southernmost of the Great Lakes and the only one whose bottom lies entirely above sea level. Its local drainage basin covers about 91,000 square kilometers (35,000 square miles), of which 23,000 square kilometers (8,750 square miles) lie north of the lake in Canada. Half of the area Erie drains is in the states of Ohio, Michigan, Indiana, Pennsylvania, and New York; the residual quarter is the lake itself. The average summer surface temperature of Lake Erie is above 24 °C (75 °F), making Lake Erie one of the most biologically productive freshwater lakes in the world, but in turn making it vulnerable to eutrophication. By the 1970s, Lake Erie was considered by many to be dead, with the lake's western basin judged the most polluted part of the entire Great Lakes system (Burns 1985; Barry 1972). Efforts related to the Lake Erie cleanup evolved at two different scales: (1) the international arena in which general policy goals for the Great Lakes were articulated bi-nationally by Canada and the United States; and (2) locally among action committees and key stakeholders.

Table 10.4 International Great Lakes Agreements

Boundary Treaty of 1909	International Joint Commission created and empowered to settle water use disputes between the United States and Canada
Great Lakes Fishery Commission of 1955	Formed by United States and Canada to eradicate the sea lamprey
Great Lakes Water Quality Agreement of 1972	Established to conduct research and identify and monitor pollution
Great Lakes Water Quality Agreement of 1978	Reaffirmed commitment to pollution reduction and to maintaining the biological integrity of the lakes
Great Lakes Water Quality Agreement of 1987	Requires United States and Canada to monitor non-point pollution and contaminated ground water

Before 1970, eutrophication resulted from inputs of phosphates from both point and non-point sources that changed the lake's water to a green, murky shade with low clarity. Excess phosphorus and water warmed by summer heat left large expanses of lake surface covered with unsightly blue-green algae. When algae died and sank, they accelerated the depletion of dissolved oxygen in lake bottom water. The Maumee River in the Global Change and Local Places study area was an especially large contributor of suspended sediments and phosphorus to the western part of the lake.

Fish species such as walleye, yellow perch, and smallmouth bass – all sought by commercial and sport fishermen – began to decrease significantly in the 1960s, and undesirable, pollutant-tolerant species such as carp increased in numbers. The decline in desirable fish species threatened the economies of communities dependent on sport fishing, which accounted for an estimated US$1 billion in local expenditures annually. Improved water quality and sport fishing was judged to be essential to maintenance of a healthy economy for western Lake Erie shoreline communities and the lake's islands. By the late 1960s, the lake was approaching a stage of complete eutrophication, with little dissolved oxygen in its waters and near-extinction of both desirable and non-desirable fish species. Understanding how non-local organizations and local residents mobilized to reverse Lake Erie's pollution suggests parallel ways in which global warming problems might be addressed.

International and national initiatives

At the international scale, several key treaties and commissions marked the growing recognition of declining water quality in the region (Table 10.4). These agreements parallel such current efforts as the Kyoto Accords aimed at establishing international objectives and targets for reducing greenhouse gas emissions.

At the national scale, the Clean Water Act of 1972 fostered reductions in the amount of phosphorus entering the lake from manufacturing and sewage treatment plants. The act required plants to obtain and renew water discharge permits based on estimates of the quantity of chemicals that could be put into water sources without harming the environment.

Renewal criteria recognized improvements in technology. The act stipulated, for example, that all municipal discharges greater than 3,800 m^3 d^{-1} reduce total phosphorus load to 1 mg l^{-1} annually (Dolan 1993). That target was not reached until 1982, but a more stringent Ohio Environmental Protection Agency standard was met in 1988 and has been maintained since (Ohio Lake Erie Commission 1998). Non-point phosphorus inputs to Lake Erie have proven less tractable, since agricultural runoff is a major source of non-point phosphorus pollution. Consequently, phosphorus flows into Lake Erie decline in dry years and increase in wet years. The Maumee Remedial Action Plan has responded with actions that will reduce agricultural runoff during periods of heavy precipitation.

Local action

An important basis for these top-down international agreements was recognition that implementation of global policies inevitably relies on mobilizing local actors for specific goals. An integral provision of the Great Lakes Water Quality Agreement of 1987, for example, was the establishment of an *Area of Concern* (AOC), defined as a seriously polluted portion of the Great Lakes where standard implementations of regulatory measures had failed to restore the environment. The creation of AOCs acknowledged that government agencies alone could not clean up pollution without assistance from local communities. Ohio contains four such regions, including the Maumee Basic AOC, of which the Global Change and Local Places Northwestern Ohio Study Area is a part.

Annex II of the 1987 Water Quality Agreement mandates that a Remedial Action Plan be formulated to eliminate pollution in each AOC. Regional Action Plans are designed to avoid setting environmental groups, industries, and government agencies against one another by asking all sectors to explore environmental problems and possible ways of amelioration and remediation, without assessing blame for the existence of the problems addressed. Though coordinated by the Ohio Environmental Protection Agency, local Regional Action Plans set their own agendas and priorities based on the premise that local actors will be more committed to solving environmental problems if they play active roles in determining what is wrong and how to fix it. A key element of Regional Action Plan success is raising public awareness of environmental issues. The Maumee Regional Action Plan funded high-school students to test water quality in area streams, painted fish on drainage inlets and basins, and conducted public forums on environmental issues. The forums were also used to recruit new members for action committees. Action committee membership is open to anyone concerned with environmental issues.

The Maumee River drainage basin deposited approximately 42% of all phosphates entering Lake Erie, with non-point sources accounting for most (Toledo Metropolitan Area Council of Governments 1991). The initial goal of the Maumee AOC was to reduce the tons of phosphorus and sediments entering Lake Erie largely from non-point agricultural sources. The Maumee River Remedial Action Plan Advisory Committee's (1990) report identified ten resulting lake and water use impairments:

• restricted fish and wildlife consumption;
• degraded fish and wildlife populations;

- fish tumors or other deformities;
- degraded benthos;
- restricted dredging;
- eutrophication or undesirable algae;
- restricted drinking water consumption or degraded taste and odor;
- beach closings;
- degraded esthetics; and
- loss of fish and wildlife habitat.

A second stage of the report proposed returning Lake Erie waters to conditions safe for swimming and conducive to fish and wildlife growth. Action groups formed in 1992 took responsibility for remediating such impairments. Five specific action plans for the Maumee River Regional Action Plan addressed pollution sources with proposed solutions rooted in infrastructure investments and more efficient land use. The Regional Action Plan's phosphorus reduction program and the spread of zebra mussels have led to improved water clarity in Lake Erie. (Though their introduction has had many negative consequences, zebra mussels do remove impurities in large quantities.)

High concentrations of toxic chemicals exist in the sediments of the Maumee Area Of Concern, which result in the degradation of benthic organisms. Biological magnification of leachates up the food chain and bioaccumulation of toxins produces impaired wildlife. PCBs in the Maumee Area Of Concern, for example, have been linked to deformities in eagles. The Maumee Regional Action Plan and the Ohio Environmental Protection Agency are now collaborating to reduce and eventually eliminate toxic leachates from landfills. The consequent improvement in Lake Erie wildlife habitat is evident in the local bald eagle population. The number of bald eagle nests in the Ohio waters of Lake Erie increased from four in 1976 to thirty-nine in 1997 (Ohio Lake Erie Commission 1998).

The Lake Erie cleanup and global warming mitigation

The organizational structures that fostered the Lake Erie cleanup might be employed to mitigate global warming (Table 10.5). Parallel structures exist whenever international agreements and national legislation provide top-down oversight policies coupled with bottom-up implementation rooted in local volunteers. There is one major difference between the Lake Erie cleanup and greenhouse gas warming, however: the lake's imminent demise was clearly evident in ways that the consequences of global warming are not. Moreover, in addition to seeing the reality of Lake Erie's deterioration, its economic effects on commercial and sport fishing were clear. The consequences of greenhouse gas emissions are longer-term and less conclusive. Cleaning up Lake Erie and future mitigation of global warming will depend on changing human behavior and attitudes toward the environment. Those changes are more easily induced when the consequences of problems and the effects of remediation are highly visible and when individuals have direct stakes in those outcomes.

The Lake Erie cleanup suggests that public education may prove to be the most effective method of overcoming apathy about global warming. Public meetings devoted to positive

Table 10.5 Comparison between the cleanup of Lake Erie and greenhouse gas policy

Pollution of Lake Erie	Greenhouse Gas and Global Warming
1. Pollution effect immediate	Pollution effect long-term
2. Negatively affected plant and wildlife	Effect on plant and wildlife unknown
3. Negatively affected public recreation	Effect on public recreation unknown
4. Scientific studies conclusive	Scientific studies inconclusive
5. Pollution hurt region's economy	Pollution effect on region unknown
6. International treaties signed	International treaties pending
7. Federal laws enacted to reduce pollution	Federal laws to reduce greenhouse gas emissions not yet enacted
8. Industrial response result of public pressure	Industrial response result of technological advances to improve fuel efficiency
9. Public able to determine behavior modification	Public behavior modification dependent on technological advances
10. Less intrusive behavior modifications required	More intrusive behavior modification required

and negative consequences of global warming could be used to recruit environmentalists and to develop local programs for reducing the negative effects of greenhouse gas emissions. As awareness of threats to local ecosystems increases, local individuals, firms, and institutions will become more receptive to programs designed to reduce greenhouse gas emissions. Though 30 years elapsed between widespread recognition of Lake Erie's condition and real improvements being evident, the cleanup demonstrates that the changes needed to reduce environmental deterioration can be accomplished.

Coal in Central Pennsylvania

Central Pennsylvania supplied coal to the country throughout the late nineteenth and early twentieth centuries. Coal consumption continues to increase in the United States overall, but Pennsylvania production dropped in the late twentieth century. The Clean Air Acts of 1970 restricted sulfur dioxide emissions and resulted in decreased demand for central and western Pennsylvania's high-sulfur coal (Munton 1998), which in turn led to unemployment and lower levels of welfare in Pennsylvania's coal-producing areas. Those linked declines illustrate the ways blanket policies can affect areas dependent upon certain kinds of production. By extension, the example emphasizes that broad prescriptions for changing existing practices formulated in response to global warming should be applied carefully and with due consideration for their economic and social consequences.[2]

---

[2] The following discussion is modified from Denny (1999), using data from County Business Patterns (United States Bureau of the Census 1964–67, 1969–1995); United States Department of Energy, Energy Information Administration (1998); and USA Counties Data Base (United States Bureau of the Census 1998).

Mining operations employ a small percentage of the national labor force, but in the five-county Central Pennsylvania study area mining employment was still 6% in 1970, and as high as 15% in Clearfield County. Although it has declined since, mining employed four to five percent of the study area labor force in the mid-1990s, again with higher percentages concentrated in the western part of the region. Declining mining employment in the study area led to unemployment levels consistently higher than those in the United States or for all of Pennsylvania. Despite the declines, the study area relies disproportionately on coal and coal mining for its economic well-being. A carbon dioxide mitigation policy curtailing coal use would bring additional pressure to bear upon an already depressed region, just as it has in the case of acid rain mitigation through sulfur dioxide reduction. Sulfur dioxide and carbon dioxide are both airborne gases directly associated with the combustion of coal. Moreover, the manner in which sulfur dioxide emissions are controlled (tradable permits) is similar to one of the strategies being proposed to control carbon dioxide emissions from fossil fuel combustion (Rose and Tietenberg 1993).

In light of these facts, differences in economic structure and change within the coal industry between Pennsylvania and Wyoming can be used to highlight the impact that sulfur regulations had on the economy of the local area, beginning with the National Source Performance Standards (NSPS) sulfur regulations of 1979 and continuing with the 1990 amendments to the Clean Air Acts that established the tradable permit system. Although low-carbon-dioxide coal does not exist, Wyoming's low-sulfur coal could be analogous to other fossil fuels such as petroleum or methane with lower carbon dioxide emissions per unit of energy. Thus the analogy illustrates how a local economy might be affected by shifts in electricity production to areas or sectors using other fuel sources.

Given the strict legislation regarding sulfur dioxide emissions, logic would suggest that the export base of the coal industry has been shifting over the past 20 years owing to changes in sulfur dioxide regulations. Indeed, local employment in the coal-mining industry has been declining. The hypothesis that this decline is related to sulfur dioxide regulation appears to be confirmed by the results of a dynamic shift-share analysis (Banff and Knight 1988) used to analyze regional employment growth over a given period and to identify anomalous years. The analysis showed that Pennsylvania coal mining has declined since the introduction of sulfur dioxide regulation; in almost every year from 1990 to 1998, coal mining jobs have been lost in Pennsylvania, while Wyoming has simultaneously seen an increase in mining jobs. In total, Pennsylvania lost more than 10,000 coal mining jobs due to competitive differences while Wyoming gained over 4,000 jobs during the same period.

Still, factors other than limits on sulfur dioxide emissions contributed to the decline in Pennsylvania mining jobs. Part of the job loss results from the general decrease in coal mining employment resulting from improved technology that has multiplied the productivity of each miner. Second, population and industry have been growing rapidly in the western parts of the United States, so there are now bigger markets for western coal. Third, Pennsylvania's bituminous coal cost about US$26.00 per ton to mine in the mid-1990s, whereas Wyoming's surface mined sub-bituminous and lignite cost only US$6.00 per ton (United States Department of Energy, Energy Information Administration 1998). More lignite or sub-bituminous than bituminous coal must be mined to produce equivalent energy, of course, but even after adjusting for energy yield, bituminous coal still costs about twice as much per unit of energy as Wyoming coal. Taking all these factors into account, the fact remains that some of the study region's employment decreases are attributable to sulfur dioxide abatement legislation. In the low-sulfur coal region of southern West Virginia, where bituminous mining costs are not radically different from those in Pennsylvania, mining and employment have increased markedly.

The sulfur dioxide – carbon dioxide analogy

Strong similarities exist between sulfur dioxide mitigation and potential carbon dioxide mitigation in the five-county Central Pennsylvania study area. In both cases, policy makers are or could be responding to a perceived threat to the biophysical environment. In both cases, residents have and would have no control over the decisions. In both cases, the negative local socioeconomic impacts are and would be serious.

The example of sulfur dioxide mitigation suggests that legislation passed to limit carbon dioxide emissions from coal combustion would affect the national coal industry in ways similar to those observed in the Central Pennsylvania study area in recent decades, and that the effects would be similar whether tradable permits or coal taxes (Muller 1996) were employed. If the Global Change and Local Places five-county region is an accurate gauge of the socioeconomic consequences of such policies, the effects would be catastrophic in some areas. Even with only 5% of the labor force directly involved in coal mining in the study area, additional people are employed in many mining-related services or in services that support miners and their families. Additional study area residents rely on coal for a cheap, readily available source of energy in their homes or workplaces. Although the overall regional economy may not suffer from the kind of changes wrought by sulfur mitigation legislation (Kamat *et al.* 1999), pockets of distress appear to be inevitable. Without a fine-meshed socioeconomic safety net, blanket prescriptions would invariably wreak ruin in selected locations. Therefore, policy makers who wish to create equitable legislation must consider its impacts on localities. One solution might be a domestic form of joint implementation (Harvey and Bush 1997) that pairs coal-producing areas with regions that generate little carbon dioxide in energy production, such as the Pacific Northwest.

Learning about mitigation from analogs

The analogs discussed for three of the Global Change and Local Places regions – Southwestern Kansas, Northwestern Ohio, and Northwestern North Carolina – provide useful insights for issues related to greenhouse gas mitigation (Table 10.6). In all three cases, local activity contributes to larger-scale problems. In all three, residents acknowledged the existence of those problems and, in North Carolina, initiated steps to mitigate them. Local mitigation efforts yielded some benefits in all cases, but could not by themselves solve wider regional problems. In all three instances, local adaptation was possible (digging deeper wells, avoiding Lake Erie, treating local water supplies), but in none of the three were the larger, enduring issues addressed. Wider voluntary concern was evident, as

Table 10.6 Global Change and Local Places analogs to greenhouse gas mitigation

Analog Characteristics	Southwestern Kansas Ogallala Aquifer	Northwestern North Carolina Watershed Protection	Northwestern Ohio Lake Erie Cleanup	Central Pennsylvania High-Sulfur Coal
Local action – larger problem	Y	Y	Y	Y
Early local concern	Y	Y	Y	N
Local mitigation, some impact	Y	Y	Y	N
Local mitigation, some benefit	Y	Y	Y	N
Local adaptation possible	Y	Y	Y	Y
Wider voluntary concern	Y	Y	Y	Y
Wider regulatory action	Y	Y	Y	Y
Local response to wider action	Y	Y	Y	Y
Wider, local benefit from action	Y	Y	Y	N
Winners and losers	Y	Y	Y	Y

was wider regulatory action at regional, state, federal, and international scales. Local citizens helped implement mitigation measures and eventually benefitted from them. The Ohio example comes closest to a win–win outcome; the detergent industry's lobbying against phosphate controls failed (Kehoe 1992). In Kansas and North Carolina, some residents will sustain costs not compensated by the larger community.

The Central Pennsylvania study area differs from the other three, but can still be evaluated within the same framework (Table 10.6). Local mining contributed to a regional and national sulfur dioxide problem of which local residents were unaware. Moreover, they would have had no way of voluntarily solving acid rain problems nor would they have benefited from doing so had they been aware of the problem. Yet once action taken elsewhere (the Clean Air Act) had an impact, the local community had some, but limited, opportunity for economic change through economic restructuring. As environmental organizations and government took stronger action, the local community gained little and lost much from regulation. The Central Pennsylvania case provides the strongest analog for the prospect of winners and losers as greenhouse gas mitigation moves forward.

The Global Change and Local Places analogs point to the importance of public involvement in the implementation of solutions, including the need for education and the role of voluntary service on committees, councils, and advisory groups, as well as activity in local politics. In all cases, the solution of a widespread environmental problem eventually found its place at the scale of the locality. Scale and place do matter, whether the problem being addressed is regional or global.

REFERENCES

Allen, M. R., and J. M. Christensen. 1990. Climate change and the need for a new energy agenda. *Energy Policy*, **18** (1): 19–24.

Alston, C. 1982. Changing Currents: State Officials Turning to Problem of the Future-Water. *Greensboro Daily News*, Greensboro, NC, 20 May.

Anonymous. 1988. N. C. Environmentalists Increasingly Wield Clout. *The Charlotte Observer*, Charlotte, NC, 28 March.

Banff, R. A., and P. L. Knight. 1988. Dynamic Shift Share Analysis. *Growth and Change*, **19**, 1–10.

Barry, J. P. 1972. *The Fate of the Lakes*. Grand Rapids: Baker Books.

Brady, D. J. 1996. The Watershed Protection Approach. *Water, Science and Technology*, **33** (4–5): 17–21.

Burns, N. M. 1985. *Erie, The Lake That Survived*. Toronto: Rowman & Allanheld.

Chiotti, Q. P., and T. Johnston. 1995. Extending the Boundaries of Climate Change Research: A Discussion on Agriculture. *Journal of Rural Studies*, **11** (3): 335–50.

Clark, W., J. Jaeger, J. van Eijndhoven, and N. Dickson, eds. 2001. *Learning to Manage Global Environmental Risk: A Comparative History of Social Responses to Climate Change, Ozone Depletion and Acid Rain*. Cambridge, MA: MIT Press.

Davenport, T. E., N. J. Phillips, B. A. Kirschner, and L. T. Kirschner. 1996. The Watershed Protection Approach: A Framework for Ecosystem Protection. *Water, Science and Technology*, **33** (4–5): 23–6.

Denny, A. S. 1999. Greenhouse Gas Emissions from Coal Combustion in Central Pennsylvania: Addressing Vulnerability Through Technological Options. Unpublished M. S. thesis, Department of Geography, The Pennsylvania State University, University Park, PA.

Dilks, D. W., and P. L. Freedman. 1994. A Watershed Event in Water Quality Protection. *Water Environment and Technology*, **6** (9): 76–81.

Dolan, D. M. 1993. Point Source Loadings of Phosphorus to Lake Erie: 1986–1990. *Journal of Great Lakes Research*, **19** (2): 212–3.

Ducker, R. 1999. Constitutionality of WSWP Program affirmed in N. C. Supreme Court Decision, http://h20.enr.state.nc.us/wswp/SprCourt.html (retrieved November 13, 1999).

Easterling, W. E. 1996. Adapting North American agriculture to climate change in review. *Agricultural and Forest Meteorology*, **80** (1): 1–53.

Easterling, W. E., N. J. Rosenberg, and C. A. Jones. 1992. An Introduction to the Methodology, the Region of Study, and a Historical Analog of Climate Change. *Agricultural and Forest Meteorology*, **59**: 3–15.

Effron, S. 1988. When It Comes to Environmental Politics, Who's Leading Whom? *North Carolina Insight*, **10** (2,3): 2–9.

German, D. B., and M. K. Hoffman. 1992. North Carolinians' Concerns about the Environment. *Popular Government*, **57** (4): 15–20.

Glantz, M. H. 1988. Introduction. In M. H. Glantz, ed. *Societal Responses to Regional Climatic Change Forecasting by Analogy*: 10–18. Boulder, CO: Westview Press.

Glantz, M. H. 1991. The Use of Analogies in Forecasting Ecological and Societal Responses to Global Warming. *Environment*, **33** (5): 10–15, 27–33.

Glantz, M. H. 1992. Assessing Physical and Societal Responses to Global Warming. In J. Schmandt and J. Clarkson, eds. *The Regions And Global Warming: Impacts and Response Strategies*: 95–112. New York: Oxford University Press.

Glantz, M. H., and J. H. Ausubel. 1988. Impact Assessment by Analogy: Comparing the Impacts of the Ogallala Aquifer Depletion and CO_2-Induced Climate Change. In M. H. Glantz, ed. *Societal Responses to Regional Climatic Change: Forecasting by Analogy*: 113–42. Boulder: Westview Press.

Gray, J. 1992. History of the Development of North Carolina's Water Supply Watershed Protection Act. In D. H. Moreau, K. Watts, R. Purdy, and J. Gray, eds. *North Carolina's Water Supply Watershed Protection Act: History and Economic and Land Use Implications*: 9–68. Raleigh: Water Resources Research Institute of the University of North Carolina.

Green, D. 1973. *Land of the Underground Rain: Irrigation on the Texas High Plains, 1910–1970*. Austin: University of Texas Press.

Gutentag, E. D., F. J. Heimes, N. C. Krothe, R. R. Luckey, and J. B. Weeks. 1984. *Geohydrology of the High Plains Aquifer in Parts of Colorado, Kansas, Nebraska, New Mexico, Oklahoma, South Dakota, Texas and Wyoming*. Washington, D.C.: United States Geological Survey Professional Paper 1400-B.

Harvey, L. D. D., and E. J. Bush. 1997. Joint Implementation: an Effective Strategy for Combating Global Warming? *Environment*, **39** (8): 14–20, 36–44.

High Plains Associates, Inc. 1982. *Six-State High Plains-Ogallala Regional Resources Study*. Printed report.

Howells, D. H. 1989. *Historical Account of Public Water Supplies in North Carolina*. Raleigh: Water Resources Research Institute of the University of North Carolina.

Howells, D. H. 1990. *Quest for Clean Streams in North Carolina: An Historical Account of Stream Pollution Control in North Carolina*. Raleigh: Water Resources Research Institute of the University of North Carolina.

ICLEI (International Council for Local Environmental Initiatives). 2002. Cities for Climate Protection. http://www.iclei.org/co2/index.htm (retrieved 4 November 2002).

Kamat, R., A. Rose, and D. Abler. 1999. The Impact of a Carbon Tax on the Susquehanna River Basin Economy. *Energy Economics*, **21** (4): 363–84.

Kansas State Board of Agriculture. 1993. *Kansas Agriculture Ogallala Task Force Report*. Topeka, KS.

Kates, R. W. 2000. Cautionary Tales: Adaptation and the Global Poor. *Climatic Change*, **45** (1): 5–17.

Kehoe, T. 1992, Merchants of Pollution? The Soap and Detergent Industry and the Fight to Restore Great Lakes Water Quality, 1965–1972. *Environmental History Review*, **16** (3): 21–6.

Kerr, R. A. 1997. Greenhouse Forecasting Still Cloudy. *Science*, **276**: 1040–2.

Kerr, R. A. 1999. Acid Rain Control: Success on the Cheap. *Science*, **282**: 1024–7.

Knight, C. G., and T. J. Rickard. 1971, Perception and Ethnogeography in Southwestern Kansas. *Proceedings of the Association of American Geographers*, **3**: 96–100.

Kowalok, M. E. 1993. Common Threads: Research Lessons from Acid Rain, Ozone Depletion, and Global Warming. *Environment*, **35** (6): 12–38.

Kromm, D. E., and S. E. White. 1981. *Public Perception of Groundwater Depletion in Southwestern Kansas*. Manhattan: Kansas Water Resources Research Institute.

Kromm, D. E. and S. E. White. 1983. Irrigator Response to Groundwater Depletion in Southwestern Kansas. *Environmental Professional*, **5**: 106–15.

Kromm, D. E., and S. E. White. 1984. Adjustment Preferences to Groundwater Depletion in the American High Plains. *Geoforum*, **15** (2): 271–84.

Kromm, D. E., and S. E. White. 1985. *Conserving the Ogallala: What Next?* Manhattan: Kansas State University.

Kromm, D. E., and S. E. White. 1986. Public Preferences for Recommendations Made by the High Plains-Ogallala Aquifer Study. *Social Science Quarterly*, **67**: 841–54.

Kromm, D. E., and S. E. White. 1990. Water-Saving Practices by Irrigators in the High Plains. *Water Resources Bulletin*, **26**: 999–1012.

Kromm, D. E., and S. E. White. 1991. Reliance on Sources of Information for Water-Saving Practices by Irrigators in the High Plains of the USA. *Journal of Rural Studies*, **7** (4): 411–21.

Kromm, D., and S. White. 1992. The High Plains Ogallala Region. In D. Kromm and S. White, eds. *Groundwater Exploitation in the High Plains*: 1–27. Lawrence: University of Kansas Press.

Mather, T. 1988. North Carolina Environmentalized. *The News and Observer*, 27 March.

Maumee River Remedial Action Plan Advisory Committee. 1990. *Stage 1 Investigation Report*. Bowling Green: Ohio Environmental Protection Agency.

Meyer, W. B., K. W. Butzer, T. E. Downing, B. L. Turner II, G. W. Wenzel, and J. L. Westcoat, 1998. Reasoning by Analogy. In S. Rayner and E. L. Malone, eds. *Human Choice and Climate Change*, volume 3, *Tools for Policy Analysis*: 217–89. Columbus, OH: Battelle Press.

Moreau, D. H., M. Moubry, and D. L. Gallagher. 1988. *Watershed Protection in Western North Carolina with Special Attention to the Pigeon River Upstream of Canton*. Raleigh: Water Resources Research Institute of the University of North Carolina.

Moreau, D. H., K. Watts, R. Purdy, and J. Gray, eds. 1992. *North Carolina's Water Supply Watershed Protection Act: History and Economic and Land Use Implications*. Chapel Hill: Water Resources Research Institute of the University of North Carolina.

Muller, F. 1996. Mitigating Climate Change: the Case for Energy Taxes. *Environment*, **38** (2): 12–20, 36–43.

Munton, D. 1998. Dispelling the Myths of the Acid Rain Story. *Environment*, **40** (6): 4–7, 27–34.

North Carolina Court of Appeals. 1996. *NO.COA95-639: Town of Spruce Pine v. Avery County* (retrieved 13 November 1999).

North Carolina Department of Environment and Natural Resources. 1999. N. C. Court of Appeals Rules the WSWP Program Unconstitutional, http://h20.enr.state.nc.us/wswp/legality.html (retrieved 13 November 1999).

Ohio Lake Erie Commission. 1998. *1998 State of the Lake Report*. Columbus: State of Ohio.

Oliver, T. 1983. Environment Called Fast Growing Cause. *Durham Morning Herald*, Durham, NC, 13 October.

Rose, A. and T. Tietenberg. 1993. An International System of Tradeable CO_2 Entitlements: Implications for Economic Development. *Journal of Environment and Development*, **2**: 1–36.

Smith, R. L. 1983. Kansas Water Law and Policy. In *Water Law and Policy in the Great Plains: Proceedings 1983 Water Resources Seminar Series*: 49–72. Lincoln: Nebraska Water Resources Center.

Solomon, B. D. 1995. Global CO_2 Emissions Trading: Early Lessons from the U. S. Acid Rain Program. *Climatic Change*, **30** (2): 75–96.

Stewart, T. 1982. State Wants Active Role In Water Management. *Greensboro Daily News*, Greensboro: 24 May.

Stoloff, N. 1991. *Regulating the Environment: An Overview of Federal Environmental Laws.* Dobbs Ferry, NY: Oceana Publications, Inc.

Taylor, J. G., M. W. Downton, and T. R. Stewart. 1988. Adapting to Environmental Change: Perceptions and Farming Practices in the Ogallala Aquifer Region. In E. E. Whitehead, C. F. Hutchinson, B. N. Timmermann, and R. G. Varady, eds. *Arid Lands: Today and Tomorrow*: 665–84. Boulder: Westview Press.

Toledo Metropolitan Area Council of Governments. 1991. *Maumee River Basin Area of Concern Remedial Action Plan,* Vol. 4. *Recommendations for Implementation.* Toledo: Toledo Metropolitan Area Council of Governments.

Turner, B. L. II, R. E. Kasperson, W. B. Meyer, K. M. Dow, D. Golding, J. X. Kasperson, R. C. Mitchell, and S. J. Ratick. 1990. Two Types of Global Environmental Change: Definitional and Spatial Scale Issues in Their Human Dimensions. *Global Environmental Change*, **1**: 14–22.

United States Department of Commerce, Bureau of the Census. 1964–1967 and 1969–1995. *County Business Patterns, Pennsylvania.* Washington, D.C.: United States Government Printing Office.

United States Department of Commerce, Bureau of the Census. 1998. USA Counties Data Base. http://govinfo.library.orst.edu/usaco-stateis.html (retrieved November 1998).

United States Department of Energy, Energy Information Administration. 1998. Energy Information Administration Website. http://www.eia.doe.gov/fuelcoal.html (retrieved November 1998).

United States Environmental Protection Agency. 1995. *Watershed Protection, A Statewide Approach.* http://www.epa.gov/OWOW/watershed/state (retrieved 13 November 1999).

Warrick , R. and W. Riebsame. 1983. Societal Response to CO_2 Induced Climate Change. In R. S. Chen, E. Boulding, and S. H. Schneider, eds. *Social Science Research and Climate Change*: 20–61. Boston: Reidel Publishing Company.

Webb, W. P. 1931. *The Great Plains.* New York: Ginn and Co.

White, S. E., and D. E. Kromm. 1995. Local Groundwater Management Effectiveness in the Colorado and Kansas Ogallala Region. *Natural Resources Journal*, **35**: 275–307.

White, S. E., and D. E. Kromm. 1996. Appropriation and Water Rights Issues in the High Plains. *The Social Science Journal*, **33**: 437–50.

Wilhite, D. A. 1988. The Ogallala Aquifer and Carbon Dioxide: Are Policy Responses Applicable? In M. H. Glantz, ed. *Societal Responses to Regional Climatic Change: Forecasting by Analogy*: 353–74. Boulder: Westview Press.

Woods, J. J., J. A. Schloss, and R. W. Buddemeier. 1995. *January 1995 Kansas Water Levels and Data Related to Water-Level Changes.* Lawrence: Kansas Geological Survey, Technical Series 8.

PART FOUR

Beyond Kyoto II: greenhouse gas reduction potentials and strategies

11

Long-term potentials for reducing greenhouse gas emissions from local areas

Thomas J. Wilbanks and Robert W. Kates

Focusing on local actions to reduce greenhouse gas emissions becomes increasingly rewarding when the analysis extends beyond current attitudes and actions toward the long term. This chapter looks out to the year 2020, with attention to the intermediate Kyoto Protocol time frame of 2010 as well. Considering the possibility that a consensus will emerge in the United States that greenhouse gas emissions should be reduced, project investigators asked how local knowledge and action could make those reductions more probable and less painful.

Global and national perspectives on long-term potentials

The Global Change and Local Places project originated in the early 1990s amid growing concerns about the likelihood of global climate change, disruptive long-range impacts of climate change, and the resulting needs to reduce anthropogenic greenhouse gas emissions and to stop degrading greenhouse gas sinks. As the project proceeded, so also did major analyses of these issues at the global and national scales, providing a backdrop for the Global Change and Local Places results as well as a basis for comparing the project's local-scale perspectives with findings from analyses at grander scales.

How much greenhouse gas emission reduction is needed?

Although it is not the only greenhouse gas, carbon dioxide concentrations are commonly used as indicators for greenhouse gas concentrations more generally. In 2000, carbon dioxide constituted about 370 parts per million (ppm) of the Earth's atmosphere. Some warming and associated physical and biological impacts were already evident. According to the Third Assessment Report of the Intergovernmental Panel on Climate Change (2001,2002), atmospheric carbon dioxide concentrations will likely increase to 540–970 ppm by 2100, resulting in a globally averaged surface temperature rise of 1.4–5.8 °C (1.8–10.4 °F) between 1990 and 2100. A moderate temperature increase by 2100, equivalent to that produced by a carbon dioxide concentration of 600–700 ppm, would benefit some parts of industrial regions of the world and harm other parts of those countries, but would harm all developing regions. If temperatures rise more than moderately, all industrial and developing regions will suffer negative effects.

According to the Intergovernmental Panel on Climate Change, keeping the global average temperature increase below several degrees would require reducing the current rate of increase in global carbon dioxide emissions. A continuation of current trends combined with considerable economic growth would result in emissions exceeding the highest ranges of the Intergovernmental Panel on Climate Change projections. Limiting temperatures to an increase of several degrees would necessitate cutting the highest range of 2100 emission levels at least in half, and keeping concentrations below 550 ppm would require 2100 emission levels substantially below 1990 releases, although emissions might climb somewhat until 2040 before dropping thereafter.

About three quarters of anthropogenic releases of carbon dioxide to the atmosphere are fossil fuel emissions. Identifying the countries responsible for those emissions leads in several directions. From the mid-1700s to the present, about 55% of total global carbon emissions from the consumption of fossil fuels and cement production have come from Europe and North America. As of 2000, the United States alone accounted for 23% of the world's carbon dioxide emissions, the largest amount by far of any single country. The United States share is down from 44% in 1950 (Marland *et al.* 2000), not because its emissions have decreased, but because emissions from other areas have risen more rapidly. Other regions' emissions (especially Asia's) will continue to increase in the future.

The role of the United States in contributing to global greenhouse gas emission reductions begins with its special responsibilities as the world's largest emitter and the world's highest per capita energy consumer (Table 11.1). Historically, United States greenhouse gas emissions per capita were more than ten times those of developing countries and more than five times the global average. And while United States per capita emissions rose 5% between 1990 and 1999, overall greenhouse gas emissions rose 12% (United States Environmental Protection Agency 2001).

The United States' role is also rooted in the moral obligation of industrialized countries to take the lead in solving a problem that they themselves have helped cause and upon which their affluence has been built. An added incentive is their substantial potential to pioneer technological and policy solutions that will be widely adopted by the global community once they are shown to be effective. Note, however, that United States emissions are projected to increase rather than decrease between 2000 and 2020.

Each year the United National Framework Convention for Climate Change (FCCC) convenes a Conference of Parties to the convention to consider actions in response to concerns about climate change. In 1997, the conference of parties meeting in Kyoto, Japan, produced an agreement called the Kyoto Protocol, in which industrialized countries agreed to reduce their 2010 greenhouse gas emissions by specified percentages below their 1990 emissions. The United States' Kyoto Protocol target was 7% below its 1990 emissions. The protocol has been signed by the United States but it has not been ratified by the Congress. In April 2001 President George W. Bush announced that the United States would not adopt the protocol and its emission reduction target, but would instead seek emission reductions through voluntary actions and technological improvements. In July 2001, all other industrial countries reaffirmed their commitments to the Kyoto Protocol, and agreed on major

Table 11.1 Carbon dioxide emissions per capita, 1990–2020

Values are metric tons per person.

	1990	1999	2010	2020
United States	5.295	5.535	6.030	6.280
Industrialized Countries	3.193	3.314	3.645	3.925
Eastern Europe/FSU	3.245	1.961	2.271	2.643
Developing Countries	0.414	0.466	0.606	0.760
World	1.105	1.018	1.150	1.296

Source: United States Department of Energy (2001), Tables A10 and A16; 1990 and 1999 Actual; 2010 and 2020 Projected.

steps toward implementation of emissions reductions. Kyoto Protocol treaty provisions adopted in Marrakech in November 2001, however, reduced the emphasis on quantitative national emission reduction targets, instead detailing an international architecture for emission reduction that is likely to show regional differences in the approaches that are emphasized.

One reason given for the controversial United States decision was uncertainty about the level of greenhouse gas emission reductions required to avoid particular long-term climate change impacts, and uncertainty about how well people can adapt to climate change. Another was the rapid growth in United States greenhouse gas emissions, which increased at a rate of 1.3% per year after 1990 because of the rapid growth of the national economy, increased energy use, low fuel prices, and decreases in fuel economy in the transportation sector. Yet another reason was the relative salience of growing emissions from major developing countries, not necessarily addressed by United States emission reduction measures. With the recent confirmation of Intergovernmental Panel on Climate Change conclusions by the National Academy of Sciences (2001), however, there seems no doubt that the United States should reduce its emissions while scientific uncertainties are resolved. How much reduction will be needed is a matter of opinion rather than science, but stabilizing emissions and then reducing them below 2000 levels seems to be a modest and reasonable goal. Pushing beyond this goal toward the Kyoto Protocol ceiling seems worth considering. Estimates of the net benefit or cost to the United States economy vary from a benefit of about US$20 billion to a cost exceeding US$160 billion, depending mainly on assumptions about the rate of technological change. Even more ambitious targets are imaginable by 2020 and beyond, as most scientific appraisals envision the need for much larger reductions in the future.

Prospects for emission reduction at a national scale

A number of comprehensive efforts have been undertaken in the United States in recent years to assess prospects for greenhouse gas emission reduction and to identify the most promising technologies and policies for reducing emissions that would not do serious harm to the national economy. Especially notable is *Scenarios of United States Carbon Reductions*

(Interlaboratory Working Group 1997). Often referred to as the 'Five Lab Study' because it was carried out by five Department of Energy national laboratories, it focuses on the emission reduction effects of technology applications. *Technology Opportunities to Reduce United States Greenhouse Gas Emissions* (National Laboratory Directors 1997), referred to as the 'Eleven Lab Study,' also emphasizes technology pathways for emission reduction. *Scenarios for a Clean Energy Future* (Interlaboratory Working Group 2000) is concerned primarily with the national policy initiatives that could contribute to greenhouse gas emission reductions.

According to these reports, pathways by which the United States might achieve significant reductions in greenhouse gas emissions include:

- national policy initiatives such as appliance or vehicle efficiency standards; power plant and industrial plant emission ceilings; cross cutting taxes on carbon emissions (carbon taxes); emission permit trading; and
- technology initiatives such as technological changes or technology improvements.

Carbon taxes could take the form of higher fossil fuel prices. The strategies are, of course, interrelated: policy initiatives can induce technology changes, and technology advances can enable policy initiatives.

The *Scenarios of United States Carbon Reductions* study concluded that the Kyoto Protocol target for the United States could be achieved by requiring fossil fuel consumers to buy a permit costing US$50.00 per metric ton of carbon (a carbon tax) and by implementing other aggressive policy initiatives. The study did not examine the political feasibility of a carbon tax or consider technology pathways or policy alternatives for reaching the Kyoto objective. *Technology Opportunities to Reduce United States Greenhouse Gas Emissions* suggested the potential for technology improvements to yield emissions reductions even greater than those called for by the Kyoto protocol, but probably over a longer time period that would permit more technological improvements. *Scenarios for a Clean Energy Future* concluded that reasonable national public policies could approach the Kyoto target by 2010, but not achieve it.

To map pathways by which the United States might achieve significant reductions in greenhouse gas emissions, assuming a supportive policy environment, the *Scenarios of United States Carbon Reductions* results were combined with the findings of parallel assessments done by the National Academy of Sciences, the Office of Technology Assessment, and the Tellus Institute Interlaboratory Working Group (1997). This report assesses national technology options at two levels: (1) potential contributions to emission reductions from four major sectors (buildings, industry, transportation, and electric utilities); and (2) contributions by particular classes of technology option within each sector. The study's results are then compared with a business-as-usual scenario (likely reductions from current emission levels per unit of economic activity in the absence of new federal government emission-reduction policy initiatives) based on an average annual rate of carbon emission growth of 1.2% (Table 11.2). Considering emissions by sector and how they might be affected by broad classes of technologies, it is evident that utility supply options have been shifting since 1998 without incentives.

Table 11.2 Potential annual reductions in carbon emissions in 2010 versus a business as usual forecast

Values are millions of tons of carbon.

Technology:	Business as Usual	High Efficiency	
Incentives:	US$0 per metric ton	US$25 per metric ton	US$50 per metric ton
Sector			
Buildings			
Energy efficiency	25	42	59
Fuel cells	na	2	3
Subtotal	25	44	62
Industry			
Energy efficiency	28	44	62
Advanced turbine systems	na	5	17
Other improvments	na	5	14
Subtotal	28	54	93
Transportation			
Energy efficiency	61	74	87
Ethanol	12	14	16
Subtotal	73	88	103
Electricity Generation			
Coal plant retirement and carbon-ordered dispatching	na	25	55
Converting coal-based plants to methane	na	9	40
Cofiring coal with biomass	na	5	17
Wind	na	2	7
Extending life of existing nuclear plants	na	3	5
Hydropower expansion	na	2	4
Power plant efficiency	na	2	8
Subtotal	na	48	136
Grand total	126	234	394

Source: Brown *et al.* (1998), Table 8.

Overall, however, the *Scenarios of United States Carbon Reductions* listing of options could be considered somewhat narrow. It does not include natural gas or electricity transmission and distribution, nor does it consider alternative fuel sources for industry or changes in the geographic arrangement of activities as a way to reduce transportation system emissions. The study finds that with cost effective efficiency improvements or a relatively small carbon tax (US$25.00 per metric ton of carbon), transportation emissions reductions rank first in order of potential importance, followed in order by industry, electricity production, and buildings. A US$50.00 per ton carbon tax could take the form of higher fossil fuel prices, emission permits, or taxes on carbon emissions. Whichever mechanism or combination is employed, that level of taxation is roughly equivalent to what would be required for the United States to meet Kyoto Protocol targets. Under that condition, electricity production yields the greatest reduction, followed in order by transportation, industry, and buildings. From a national perspective, potential reductions are greater for transportation than for industry and higher for industry than for buildings when incentives are moderate, but reductions associated with electricity production exceed any of the other three when incentives are increased to US$50.00 per ton of carbon.

Within the four major sectors the most promising alternatives are:

- *Electricity generation*. Potentials for greenhouse emission reductions in this sector could be achieved mainly by retiring coal-fired plants and implementing carbon-ordered dispatching (buying lower-carbon-based electricity first). Next most important is conversion of coal-fired plants to natural gas. Co-firing coal with biomass becomes a significant alternative with the high carbon tax. By 2010, emissions would be reduced by 21% at a US$25.00 per metric ton carbon tax and by more than 50% at US$50.00 per ton, compared with a business-as-usual projection
- *Transportation*. Potentials in the transportation sector are closely tied to energy efficiency improvements rather than fuel switching. Light trucks and passenger cars would yield greater reductions than using methanol. Freight trucks are also important, as is, to a lesser degree, air transport. Reductions by 2010 compared with business as usual would be 14% with a low carbon tax and 16% at US$50.00 per metric ton.
- *Industry*. Potential reductions from energy efficiency improvements are greater than emissions reductions realized by using advanced technologies, although technology changes would become more effective if the carbon tax was higher. More reductions could be achieved in light than in heavy manufacturing because light manufacturing is projected to grow more rapidly. In heavy manufacturing, the largest potentials are in the petroleum and chemical industries, followed by iron and steel, and then in pulp and paper. Reductions in the low and high carbon tax cases would be 10 and 17%, respectively.
- *Buildings*. In residential structures the reductions could be realized by switching appliances from electricity to gas, increasing lighting efficiency, and more efficient refrigeration, space heating, and air conditioning. In commercial buildings, the greatest potentials lie in more efficient space heating and cooling, ventilation, refrigeration, lighting, and water heating, as well as in fuel switching. Estimated reductions are 8% with the US$25.00 carbon tax and 11% if the tax were doubled.

In the absence of other aggressive policy initiatives that are not assumed in the Interlaboratory Working Group (2000) scenario, only the industrial sector's reductions approximate the 1990 emission levels that constitute the Kyoto Protocol target. Even with a US$50.00 per ton carbon tax, the transportation and buildings emissions are greater (19 and 11%, respectively) than 1990 levels. Comparisons with 1990 levels were not calculated for electric utilities, but it appears that a US$50.00 per ton carbon tax would go far toward moving that sector toward its Kyoto target.

Scenarios for a Clean Future identifies realistic national *policy* alternatives that would cut across technology categories. For buildings, key policies are efficiency standards and voluntary labeling. For industry, voluntary programs are emphasized, including voluntary agreements with individual industries and trade associations. For transportation, fuel economy policies and agreements would be effective. In electricity generation, utility restructuring, renewable energy portfolio standards, and production tax credits would work best. By 2010, the most advanced policy scenario would reduce carbon emissions by 19% from business-as-usual levels, which would be only slightly less than 1997 emissions and 7% above 1990 emissions. By 2020, the most advanced case would reduce emissions to 1990 quantities or slightly less.

In summary, United States greenhouse gas emissions could be reduced by 2010, but probably not to the Kyoto target levels in the absence of more rigorous national policy intervention than has been assumed in these analyses. Further emissions reductions could be achieved by 2020, but realizing them would require policy measures and incentives that exceed those that have been deemed feasible to date.

Prospects for emission reduction at a global scale

Prospects for global emission reductions consonant with the Kyoto Protocol targets are little more promising than the probability that the United States will meet Kyoto goals. The numerous alternative paths for economic and energy resource development offer many possibilities, however, and there is at least some reason for optimism: critical technologies are changing more quickly than was earlier expected, and possible changes in institutional structures could lead to additional reductions. Attempts to assess those prospects are highly dependent on the assumptions incorporated into the scenarios used to estimate emission reductions.

The Third Assessment Report of the Intergovernmental Panel on Climate Change (2001, 2002) assesses the actions needed to achieve particular greenhouse gas concentration stabilization levels in terms of bottom line reductions in sectors rather than focusing on how the reductions are to be achieved. According to the Intergovernmental Panel on Climate Change report, emission reductions would be enabled by a number of sectoral technology and policy initiatives reviewed in Chapter 3 of its Working Group III report (Table 11.3). The specific technologies that could contribute to these emission reductions are listed in considerable detail, but contributions of individual approaches to emission reductions are generally not estimated. Potential reductions could range between 14 and 23% of the projected emissions for 2010 and 23–42% of projected emissions for 2020 – much higher than similar projections from the United States.

Overall, the Intergovernmental Panel on Climate Change analysis presents a vision of great potential, tempered by great complexity and uncertainty. Its best case scenarios are more optimistic about reducing emissions below 1990 levels in industrialized countries than are the United States analyses, except for electricity generation. Perhaps one difference is that the Intergovernmental Panel on Climate Change global assessment is farther removed from the political realities of national policy-making, which makes some of its ifs rather large. By contrast, the analyses for the United States were shaped by a review process that

Table 11.3 Estimated potential greenhouse gas emission reductions in 2010 and 2020 at the global scale

Values are million tons of carbon equivalent per year.

	Emission Share	Potential Reductions	
	1990	2010	2020
Actual or Projected Emissions:	5,281	11,500–14,000	12,000–16,000
Sector			
Buildings	1,650	700–750	1,000–1,100
Transportation	1,080	100–300	300–700
Industry	2,300	–	–
Energy efficiency	–	300–500	700–900
Material efficiency	–	c. 200	c. 600
Non-CO$_2$ gases	–	c. 100	c. 100
Agriculture	1,460–3,010	150–300	350–750
Waste	240	c. 200	c. 200
ODC Replacement	–	c. 100	n.a.
Energy supply and conversion	(1,620)[a]	50–150	350–700
Total	6,900–8,400	1,900–2,600	3,600–5,050

[a] Included in sector totals above.

Source: IPCC Working Group III, Chapter 3, Table 3/37. Reduction percentages are calculated from estimated ranges of emissions of 11,500 to 14,000 for 2010, and 12,000 to 16,000 for 2020, roughly comparable with the SRES-B2 scenario.

screened out assumptions that were deemed politically and economically unrealistic even if they were technologically possible. Another possibility, of course, is that the international community is more bullish about prospects for political will over the next two decades.

Local perspectives on long-term potentials

National and global analyses constitute a view from the top. Assessing possibilities and probabilities for reducing emissions at a local scale raises different basic issues regarding estimates and forecasts over a twenty-year time span. The Global Change and Local Places research team developed a general approach for modeling emission reduction potentials in this longer term and then applied it to the four Global Change and Local Places study areas. The results of those analyses suggest that reducing emissions will not be accomplished in the absence of stringent policies or dramatic changes in the current practices that produce greenhouse gas emissions.

Assessing emission reduction potentials for localities

The Global Change and Local Places analysis of mitigation potentials for 2010 and 2020 is grounded in the Global Change and Local Places conceptual framework (Figure 2.9).

Greenhouse gas emissions in a locality could be reduced by:

• Changing certain driving forces (recognizing that few of them are controlled within the local area);
• Reducing some major emitting activities by substituting other kinds of the same activity that do not diminish quality of life, such as cleaner high-technology industry for traditional industrial activities; or
• Reducing emissions from certain activities that are now major emitters.

To clarify the potentials for these alternatives, the Global Change and Local Places team set two arbitrary emission reduction targets, assessed alternative pathways for meeting them, and then estimated the likelihood of different causal factors that would foster those pathways. The analyses focused on whether local knowledge and opportunities for local action might make emission reductions more likely, more attractive, or less costly compared with generic emission reduction policies and actions.

Obviously, the wide range of driving forces and other causes can and probably will change over twenty years, some profoundly. Technological breakthroughs, new findings about the nature, rate, and impacts of global climate change, and acceptance of policy initiatives that now seem impracticable should all be expected. Rather than trying to forecast such changes as the possible replacement of the Kyoto Protocol with different national emission ceilings toward the end of the first decade of the twenty-first century, the Global Change and Local Places team asked what conditions might permit arbitrary local emission reduction targets to be reached. In some cases, certain conditions would be needed that do not now exist.

Modeling emission reductions

The general Global Change and Local Places approach to modeling emission reduction potentials in the four study areas (Figure 11.1) relies on three scenarios of driving forces and emission trends to 2010 and 2020 and three sets of structural assumptions, leading to mitigation actions.

For driving forces and emissions, the three postulated scenarios of 'business as usual' futures are:

• a simple *extrapolation of empirical past emission trends* from 1970 to 1990 forward to 2010 and 2020;
• emission estimates based on projections of demographic and economic conditions at county levels for 2010 and 2020 prepared by *Woods & Poole*, a commercial forecasting firm (Box 2.2); and
• the use of *best local knowledge* in forecasting greenhouse gas emissions in 2010 and 2020, based on modifications of Woods & Poole estimates to fit local expectations as to whether the most likely futures for particular sectors are best estimated by the low, medium, or high Woods & Poole forecasts for those sectors.

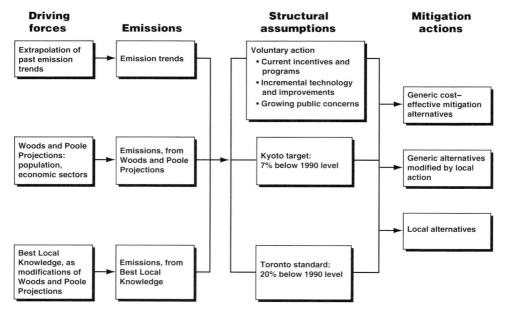

Figure 11.1 Modeling local emission reduction potentials.

For scenarios two and three, the unmodified and modified Woods & Poole projections of economic and demographic drivers were converted to emission estimates.

Estimates of local study area emissions in 2020 are compared with three structural assumptions: *voluntary action* and *Kyoto* and *Toronto* emission reduction targets:

- A reduction of 7% from the 1990 emission level, the United States emission reduction target defined in the Kyoto Protocol of the Framework Convention on Climate Change (1997). That level is the perceived need for mitigation based on perceptions of global climate change prevalent in the late 1990s; and
- A reduction of 20% below the 1990 emission level, the *Toronto Standard* adopted by the city of Toronto and many other cities in the International Council for Local Environmental Initiatives Cities for Climate Protection program (see Chapter 1). That target could emerge in the coming decade if concerns about reducing rates of greenhouse gas emission increases become more urgent.

Local greenhouse gas emission sources and sinks were then examined in the context of economic and population trends to identify three sets of mitigation actions that would reduce local emissions by 7% and by 20% below their 1990 levels. For example, in the set of *local alternatives* a fossil fueled power plant might be converted to natural gas, transportation energy efficiency might be improved by one third, or local industrial development might take different courses. These projected outcomes are based on qualitative assessments by each Global Change and Local Places study area team, drawing on information from the local study and other sources of local knowledge.

Table 11.4 Percentage reductions from best local wisdom scenarios for 2020 for each of two emission reduction targets

Percentages for the United States are 39 and 47, respectively.

	1990	2010	2020 Woods & Poole	2020 Best Local	7% Reduction	20% Reduction
Southwestern Kansas	7.62	9.3	9.98	8.17	7.09	6.10
Northwestern North Carolina	18.22	19.26	24.07	26.93	16.94	14.58
Northwestern Ohio	40.15	47.39	50.54	53.06	37.34	32.12
Central Pennsylvania	19.24	25.06	25.64	25.07	17.89	15.39

Source: EIA reference case forecasts.

This local portfolio of mitigative actions based on local data and knowledge is then compared with *generic cost-effective alternatives* that have been identified

- for the entire United States, such as the United States Department of Energy's *Scenarios of United States Carbon Reductions* (Interlaboratory Working Group 2000); and
- at global scale, such as the Intergovernmental Panel on Climate Change, to see whether assessing mitigation potentials at a local scale produces understandings different from those that emerge from assessments at national or global scale.

Finally, the portfolio of potential actions to meet emission reduction targets was assessed in terms of paths for implementation as a *generic set modified by local action alternatives.* In some cases, a particular option or two will be desirable for reasons other than greenhouse gas emission reduction alone, such as fuel cost reductions. Many of the emission reduction strategies being developed by participants in the Cities for Climate Protection program are of this type. In other instances, mitigation options will likely become viable through local agency only under certain conditions, such as growing public concern about climate change or the introduction of new emission reduction regulations or incentives at the national level. The final stage of the exercise outlines the conditions that might foster local emission reductions.

Local knowledge and opportunity

The four local case studies offer insights into the feasibility of greenhouse gas emission reduction through local agency in the United States, the levels of costs and other difficulties involved, and the incentives that might be required in order to reach different goals. Table 11.4 presents estimated emissions for 1990 and the targeted reductions of 7% and 20% compared with projected emissions for 2010 and 2020 using the Woods & Poole population and economic projections, and as modified for 2020 by the best local knowledge. In general, comparing the use of local knowledge about trends with the more generalized Woods & Poole projections gives larger emissions in 2020 for all sites except Southwestern

Table 11.5 Percentage reduction required from 2020 emissions to meet 7% and 20% targets in Global Change and Local Places study areas and in the United States

	7% less than 1990	20% less than 1990
Southwestern Kansas	13	25
Northwestern North Carolina	37	46
Northwestern Ohio	30	39
Central Pennsylvania	29	39
United States[a]	39	47

[a] Based on EIA reference case forecast.

Kansas. By 2020, in order to meet the 7% Kyoto target, reductions in the best local knowledge emissions projection would need to range from 13% in Southwestern Kansas to as much a 37% in Northwestern North Carolina. (Table 11.5). For the 20% Toronto target, reductions would range between 25% in Southwestern Kansas and 46% in Northwestern North Carolina. Except for Southwestern Kansas, the study areas are relatively consistent as regards the projected impacts of emission reduction targets: roughly 30–35% for the Kyoto 7% reduction from 1990 levels, and roughly 40–45% for a 20% Toronto reduction. The difference between a 7% target reduction and a 20% target is not enormous in terms of the impact on a local area; the dominant factor is using the 1990 base rather than a later year.

Reducing greenhouse gas emissions between one third and one half from the levels projected to exist in 2020 would, of course, be a non-trivial undertaking. In most cases, actions would have to be taken by the major emitters in each area or by the major emitting sectors, although a range of smaller savings would certainly ease the pressure somewhat. According to assessments by the study teams based on local knowledge, the major pathways toward emission reductions of this magnitude would call for the kinds of actions outlined below for each study area. These local assessments are broad qualitative judgments, generally not buttressed by quantitative estimates of how much particular pathways could contribute; but they are listed in the approximate order of the importance of each potential reduction.

Southwestern Kansas

Based on the local assessment, the quick way to reduce local greenhouse gas emissions would be to shut down the Holcomb Station electric power plant, which would reduce total emissions by 2.2 million short tons of carbon dioxide equivalent – enough to meet either a 7% Kyoto Protocol or a 20% Toronto Standard emission reduction from 1990 emission levels in 2020. Obviously, the electricity generated at Holcomb would have to be supplied from another source, which might simply shift the emissions outside the study area. An alternative would be to power the plant with natural gas instead of coal, which could cut power plant emissions by nearly one half. Other pathways that would reduce emissions from the area include:

- Reducing local emissions from natural gas production and transmission, for example by increasing the efficiency of natural gas compressors and shutting down wells with low wellhead pressures. A reduction in emissions of one fourth from 1990 levels would mean about 0.5 million tons of carbon dioxide equivalent. Of course, if the country as a whole increases its use of natural gas as the lowest impact fossil fuel, such local initiatives could work at cross purposes to larger scale initiatives with the same overall goals.
- Improving the efficiency of truck transportation, which is a major emitter in the region, although efficiency improvements will be offset by increased transport if the agricultural processing sector continues to expand locally. If truck freight miles were to remain unchanged from 1990 to 2020, a 20% improvement in truck efficiency would reduce emissions by about 0.2 million tons of carbon dioxide equivalent.
- Reducing emissions from integrated agricultural and animal production, for example by changing feed mixes to reduce methane emissions from cattle feedlots. Potential savings via this path are probably not large compared with power plant and natural gas compressor emissions.

Without a strategy focused on the energy sector, emission reduction targets probably cannot be met in Southwestern Kansas unless the demand for meat products decreases or some of the agricultural processors move out of the area. Clearly, local residents would rather adapt to climate change than experience either of these possibilities.

Northwestern North Carolina

As in Southwestern Kansas, the simplest and most direct way to reduce local greenhouse gas emissions would be to shut down the two coal-fired power plants or to convert them to natural gas. Eliminating their current emissions would reduce total emissions by more than eight tons of carbon dioxide equivalent, a majority of the emission reductions needed to meet the hypothetical emission reduction targets of ten tons for the Kyoto 7% criterion and 12.5 tons for the International Council for Local Environmental Initiatives 20% goal. Other potentially rewarding pathways include:

- Improving transportation energy efficiency, especially for private vehicles. Changes in the geography of jobs and residential locations could make a difference, but they are unlikely. Even if vehicle miles were to remain unchanged, however, a 20% improvement in efficiency would reduce emissions by less than 1.0 million tons of carbon dioxide equivalent.
- Improving industrial fuel efficiency by replacing inefficient older factories with newer, more efficient plants or by accelerating the trend toward service employment and out of such traditional forms of industrial production as textiles and furniture. Increasing the substitution of local biomass for fossil energy sources used in manufacturing would also help. Again however, even if industrial capacity were to remain unchanged rather than growing at the nearly 50% rate incorporated into the 2020 projection, a 40% efficiency

improvement would reduce emissions less than one million tons of carbon dioxide equivalent. It seems more likely that efficiency improvements would keep industrial emissions from climbing as the economy expands. Similarly, a very high rate of fuel substitution would be required to keep pace with projected industrial growth.

• Decoupling residential quality of life from increased energy consumption, which would imply cultural change as well as technological change.

Because the Northwestern North Carolina study area is expected to grow more rapidly than the other three study areas between 1990 and 2020, its emission reduction targets are the most stringent. They appear to be impossible to meet without terminating emissions from local power plants, which seems exceedingly unlikely and to a large degree shifts emissions elsewhere outside the study area. Even assuming that local power plant emissions could be eliminated and that substantial reductions in transportation and industry emissions could be achieved during a period in which both sectors will be expanding, the 7% Kyoto Protocol target appears to be impossible to meet and the 20% Toronto standard goal lies well beyond any strategy other than a major social and economic restructuring of the region.

Northwestern Ohio

The Northwestern Ohio study identified three pathways for local initiatives to reduce greenhouse gas emissions in this area:

• Energy efficiency improvements in heavy industry;
• Energy efficiency improvements in the study area's transportation system; and
• Land-use planning and urban development planning to reduce urban sprawl.

Assuming a growth rate of at least 20% for industry overall, as implied by the 2020 projections, a net efficiency improvement of 40% would reduce Northwestern Ohio's 2020 emissions by less than 3.5 million tons of carbon dioxide equivalent, compared with targets of 15.7 (Kyoto Protocol) and 20.9 million tons (Toronto standard). If land-use planning and vehicle efficiency improvements combined to reduce total transportation emissions by 30% while the economy is growing, the net emission reduction would be about 3.0 million tons of carbon dioxide equivalent. Nearly 5.0 million tons of carbon dioxide equivalents could be reduced by eliminating all electric power production in the study area except that produced by nuclear energy, and another 1.0 or 2.0 million ton reduction might be realized in the residential and commercial sectors. Even so, total reductions still fall far short of the Kyoto and Toronto targets. Even a 7% Kyoto target could be met only by restructuring the study area's heavy industry economy, replacing heavy industry with economic activities that emit fewer emissions per unit of output.

Central Pennsylvania

As in Southwestern Kansas and Northwestern North Carolina, the best prospects for reducing greenhouse gas emissions center on reducing emissions from coal-fired electric power plants. For this purpose in Central Pennsylvania, technologies that increase the efficiency of coal combustion are far more likely to find acceptance than switching to other fuels. No other general pathway appears to have a chance of making a significant difference in emissions other than shutting down or reducing the production of electricity and lime in the study area, neither of which appears probable. If, while the economy is growing by one third, coal use were to remain constant and greenhouse gas emissions were reduced by one third (big ifs indeed), actual emissions in 2020 would be down by about 5.0 millions tons of carbon dioxide equivalent from projections, compared to target reductions of 7.0 million tons to meet the Kyoto standard and almost 10.0 million tons for the Toronto standard. Emissions from all other sectors would have to be reduced an average of 20% in addition to the coal use efficiency improvement to meet even the 7% Kyoto target.

In general, the Global Change and Local Places local study teams estimate that achieving a reduction of greenhouse gas emissions in the study area by 30–40% below anticipated 2020 levels would require dramatic, discontinuous changes at the largest point sources of emissions. In three of the four study areas, this means coal-fired electric power plants. In Northwestern Ohio, it means large, energy intensive industrial plants. Short of such unlikely changes, ambitious targets probably cannot be achieved. In three of the four study areas, transportation system efficiency improvement was also identified as a possibility for major emission savings. In none of the four study areas were the residential or commercial sectors seen as candidates for substantial emissions reductions, except in relation to demand for transportation with respect to urban sprawl in Northwestern Ohio and the typical journey-to-work distances in Northwestern North Carolina.

An alternative view of local emission reduction potentials

Besides the best local knowledge approach utilized by Global Change and Local Places researchers to estimate potentials for emission reductions by 2020, other approaches may lead to different answers. One example is the analyses developed by the Cities for Climate Protection Campaign of the International Council for Local Environmental Initiatives, now being used in more than 300 cities and localities worldwide to identify end-use emission reduction potentials (Chapter 1, p. 20). Unlike the Global Change and Local Places approach, the International Council for Local Environmental Initiatives (ICLEI) analysis does not ask whether emission reductions are *likely*; rather, it asks how they could be achieved with known or plausible technologies. With the assistance of Ralph Torrie and others from the ICLEI, its methods were applied to the Northwestern North Carolina and Northwestern Ohio study areas. The results vary from some of the local judgments.

In Northwestern North Carolina, for example, the most obvious difference, aside from the emphasis local experts placed on power plant emissions that were not considered in the end-use-oriented ICLEI analysis, is the importance in ICLEI results of emission reduction potentials in the residential and commercial sectors (Table 11.6). The potential reductions suggested by ICLEI protocols for the transportation and waste sectors are also notably higher than local wisdom estimates. In Northwestern Ohio (Table 11.7) the most obvious difference

Table 11.6 Potential for reducing emissions 20% below 1990 levels in the Northwestern North Carolina study area

Percentages are the proportions of the study area's emission reduction targets achieved by that sector.

Sector	Local Best Wisdom	ICLEI Analysis
Residential	Little potential	28% of target achievable through such measures as a 50% increase in appliance and lighting efficiency and a 25% increase in building envelope efficiency
Commercial	Little potential	19% of target possible through such measures as a 50% increase in lighting and equipment efficiencies
Industrial	Will do well to keep total emissions from increasing as economy expands	15% of target possible through such measures as a 20% improvement in lighting and equipment efficiency
Energy	Up to 65% of target possible	Not explicitly considered
Transportation	Less than 10% under the most optimistic conditions	29% of target possible through such measures as a 50% increase in highway vehicle fuel efficiency
Waste and Other	Not explicitly considered	10% of target possible through such measures as landfill methane gas recovery

between local wisdom and ICLEI analysis is that the ICLEI end-use view identifies potentials that exceed by one third the 20% reduction target, leading to optimism that that goal could be reached. Again, power plant emissions make a difference, but the ICLEI methodology also identifies significant potential reductions in the residential and waste sectors and much larger potentials in the industrial and transportation sectors.

These contrasts suggest three possible general explanations for the differences between the estimates produced by the Global Change and Local Places project and the International Council for Local Environmental Initiatives analysis, all of which may contribute to the differences. First, as noted above, the ICLEI analysis does not consider whether its measures are *likely* to happen under realistic local circumstances; it simply asks whether such measures *could* achieve the savings under realistic assumptions about technology and possible technological changes. Second, Global Change and Local Places local experts concentrated on measures under the control of local agents or at least related to local processes, which tended to limit their attention to technological changes resulting from national and international policies and market transformations, though an advocate for the ICLEI approach might contend that power plant emissions are not under the control of local agents, especially in city government. Third, local experts may have underestimated potentials for emission reductions through these larger forces, especially in the residential, commercial, and transportation sectors. If these interpretations are correct, the prospects for meeting a

Table 11.7 Potential for reducing emissions 20% below 1990 levels in the
Northwestern Ohio study area

Percentages are the proportion of the study area's emission reduction target achieved
by that sector.

Sector	Local Best Wisdom	ICLEI Analysis
Residential	Perhaps 5–10% of target possible	27% of target achievable through such measures as a 50% increase in appliance and lighting efficiency and a 25% increase in building envelope efficiency
Commercial	Perhaps 5–10% of target possible	10% of target possible through such measures as a 50% increase in lighting and equipment efficiencies and a 25% increase in building envelope efficiency
Industrial	Perhaps 15% of target possible	26% of target possible through such measures as converting fuel use from coal to natural gas and a 20% improvement in lighting and equipment efficiency
Energy	Up to 25% of target possible	Not explicitly considered
Transportation	Perhaps 15% of target possible	42% of target possible through such measures as a 50% increase in highway vehicle fuel efficiency and conversion of 20% of auto fleet to gasoline–electric hybrids
Waste and Other	Not explicitly considered	34% of target possible through such measures as landfill methane gas recovery and a 50% reduction in emissions

20% emission reduction target may not be quite as unlikely as the local analyses suggest,
though they still appear daunting.

Scale and differences in emissions reduction opportunities

It seems possible that in estimating potentials for greenhouse gas emission reductions, scale
differences lead to consistent biases in both optimistic and pessimistic directions. Assessments by external experts based on national or international scale pathways are likely to err
on the side of optimism because the outsiders fail to appreciate local market realities and
barriers. At the same time, it is possible that assessments done by local experts may underestimate emission reduction potentials because local residents fail to appreciate market and

Table 11.8 Relative contributions of sectors to greenhouse gas emission reductions for 2020

Assessment and Scale:	GCLP (Local)	ICLEI (Local)	DOE (National)	IPCC (Global)
Sector				
Buildings	Small	Large	Medium to Small	Large
Industry	Small to Large	Medium to Large	Medium	Medium to Large
Transportation	Medium	Medium to Large	Medium	Medium
Agriculture	Large to Negligible	Not Estimated	Not Estimated	Medium
Waste	Not Estimated	Small to Large	Not Estimated	Small
Energy	Large	Not Estimated	Large to Medium	Medium

technology trends operating at broader scales whose effects will trickle down to their areas without local agency or choice. If so, assessments should more often involve interactions between experts at both scales in order to examine and possibly balance limitations of both.

But does the local scale perspective identify possibilities that are overlooked in broader scale analysis? Assessments conducted by the Global Change and Local Places team, the United States Department of Energy, and the International Panel on Climate Change all tend to be highly generic, focusing on broad sectors rather than specific technologies. Consequently, many possible differences (if they exist) may be obscured by aggregation.

Yet another possibility is that differences among assessments of greenhouse gas emission reduction potentials may be shaped less by scale of analysis than by the perspectives of analysts whose views are inevitably conditioned by institutional agendas, personal interests, and the aspects they choose to emphasize or ignore. When assessments made across a range of scales from local through national to global are compared, it is clear that different relative weights are assigned to different sectors, and certain sectors are ignored in some studies (Table 11.8). The local scale Global Change and Local Places and national Department of Energy analyses, for example, show similar reduction potentials for the buildings, energy, and waste sectors, in contrast to the local scale International Council for Local Environmental Initiatives or the global scale Intergovernmental Panel on Climate Change assessments. Aside from these differences, however, the general perspectives are relatively consistent assessments. Energy, industry, and transportation are estimated to be moderate or large contributors when they are considered, not surprisingly since these are known to be the dominant fossil fuel users and greenhouse gas emitters. Waste and agriculture are usually estimated to have low emission reduction potentials if they are considered at all. The largest difference among the assessments is for potential reductions in the buildings sector, which are estimated to be large in the local scale International Council for Local Environmental Initiatives study and global scale Intergovernmental Panel on Climate Change assessment, but smaller in the national scale Department of Energy and Global Change and Local Places analyses.

The insights offered by local knowledge become more evident at a lower level of generalization: i.e. in the details. Instances from the Global Change and Local Places project are particular driving forces, distinctive emission reduction strategies, and perverse policy and technology impacts. A specific example is the importance of the evolving geographies of local economies in the Northwestern Ohio and Northwestern North Carolina study areas, which suggest policies that might address the long average distances between home and work. Distinctive emission reduction strategies arise from using biomass instead of fossil fuel in industry in Northwestern North Carolina and from the roles played by transnational corporate policies within the refinery sector in Northwestern Ohio. A complex interplay of local and national policies is evident in the tradeoffs between local emission reductions and national reductions with respect to natural gas transmission in Southwestern Kansas, the negative local impacts of national policy on the use of high-sulfur coal in Central Pennsylvania, and the perverse power plant fuel regulation that led to siting a coal-burning electric power plant atop a major natural gas field in Southwestern Kansas. Often hidden by sectoral totals, these details can be critically important in designing effective emission reduction strategies at any scale

Perspectives on prospects for reducing greenhouse gas emissions through local knowledge and opportunity

The Department of Energy studies summarized above were prepared by scientists and engineers who were optimistic about technological fixes but cautious about assuming forceful policy interventions. Those analysts concluded that an emission reduction target such as the Kyoto Protocol could be achieved with existing technology, but only if policy intervention was more stringent than now appears to be feasible politically. The challenge facing the United States is made more difficult by the consistent growth in emissions since 1990 during a time when some other industrialized countries have stabilized or even reduced their emissions (even China has decreased its emissions in recent years). Rather than imposing stringent regulations that would reduce greenhouse gas emissions, the United States has instead proposed voluntary action and technology improvements whose effects will be evident only after 2010. These developments suggest that potentials for substantial reductions in greenhouse gas emissions in the United States in the near term will depend mainly on decentralized actions such as the state of Oregon's prohibition on net increases of greenhouse gas emissions from new electric power plants in the state and programs undertaken by cities in the International Council for Local Environmental Initiatives Cities for Climate Protection network (Chapter 1, p. 20). While encouraging, these initiatives represent only a modest share of state, local, and corporate entities, perhaps no more than 10%.

Given what has been learned from the Global Change and Local Places analyses of its four study areas, how likely is it that local agency will contribute to substantial greenhouse gas emission reductions such as the Kyoto Protocol targets? Global Change and Local Places results are sobering about prospects for local action in a geographic sense. Under conditions that the local research teams and local experts consider realistic, meeting commonly mentioned emission reduction targets in the four study areas will be virtually impossible without either shifting the emissions to other areas or accepting adverse local economic consequences. To a considerable degree, emission reductions as results of

local agency are unlikely because so many of the key pathways involve structures, organizational decisions, and technologies originating from or controlled elsewhere. In fact,
a focus on local emission reduction is likely to provoke local opposition unless citizens
become convinced that global climate change is a problem about which they should do
something. Local opposition is likely to intensify if global climate-change-related policies
are implemented that are insensitive to local conditions, and especially if they are invoked
without consulting local leaders. It is possible that prospects for decentralized action by
corporations are somewhat brighter, because their ability to make and implement significant decisions is more focused if broader corporate policy is supportive. An overriding
question is whether policy, technology, or social conditions will change sufficiently before
2020 to make the posited emission targets achievable through local agency, including local
advocacy for action by translocal institutions.

On the basis of Global Change and Local Places experience to date it appears that local
actions to reduce greenhouse gas emissions from localities will require:

- Convincing evidence of disruptive effects of global climate change;
- Documentation that emission reductions will insulate localities from costs associated with
  global climate change or offer opportunities or other benefits due to climate change;
- Incentives and assistance for local innovation, related to a degree of local participation in
  designing and implementing strategies; and
- Technology improvements to reduce emissions that are appropriate to local conditions
  and that do not substantially increase costs or inconvenience.

If these four conditions were to be met, local action could make a real difference, and the
International Council for Local Environmental Initiatives perspective suggests that they
are not unimaginable. For example, the current United States assessment *The Potential
Consequences of Climate Variability and Change* (United States Global Change Research
Program 2000) raises generic issues for local sites such as increased drying of agricultural
and ranch lands in Southwestern Kansas or lowered Great Lakes levels for Northwestern
Ohio. Assessments that would include locally analyzed vulnerabilities as well as local measures of the many physical and biological changes currently underway would raise local
interest and increase local attention to the disruptive impacts of global climate change. The
experience of the Cities for Climate Protection campaign argues for coupling local concerns about air pollution, energy costs, traffic, and sprawl with net benefits of greenhouse gas
emission reduction for economic and environmental conditions in a local area. Incentives
and assistance for such local or regional opportunities as biomass burning in Northwestern
North Carolina or feedlot diet improvement in Southwestern Kansas can encourage local
innovation. Finally, technology research, development, and adaptation to expand the portfolios of alternatives available in addressing locally varying market conditions can shift
local willingness to address climate change issues, as exemplified by the political interest in Southwestern Kansas in the potential for sequestering carbon in prairie and plains
soils.

The answer to the question posed for this chapter, then, is that the potential to reduce
greenhouse gas emissions in the United States through local action is limited if local areas

must depend on local agency without encouragement and assistance from external sources. The potential for reducing emissions could be very bright, however, if local agency can be connected to external parties in effective macro–micro-scale partnerships.

Local analysis and national policy

What will be lost if policy-making designed to reduce United States greenhouse gas emissions proceeds in the absence of analyses conducted at locality scale? The answer is subtle but significant. Local-scale analyses are not likely to change the bottom line for emissions at broad scales of generalization, but local analyses permit policymakers to appreciate and take advantage of the geographical texture of effective responses to concerns about climate change issues. Global Change and Local Places local-scale analyses identified pathways for emission reduction that have been overlooked in aggregate analyses, such as improved natural gas distribution system efficiency, using biomass energy to displace coal, urban morphology as a determinant of transportation energy use, of the viability of such energy-efficient alternatives as small-scale combined heating, cooling, and power generation facilities. Local analyses also expose the weaknesses and unintended consequences of 'one size fits all' national policy proposals. If known beforehand, the insights resulting from local-scale analyses can lead to more effective and more equitable national policies. In addition, local-scale analyses could detect changes in public attitudes and consumption patterns that occur as forms of autonomous adaptation to climate change and other developments. Capturing the existence and extent of such adaptations would be invaluable in formulating scenarios – and policies – for the longer term.

Finally, in a democratic society such as the United States, local acceptance and compliance with mitigation policies is a prerequisite to the success of those measures. Americans have proved adept at evading regulations and policies with which they do not fundamentally agree, and they will ignore or work around top-down regulations that fail to take salient local conditions into account. The results of the Global Change and Local Places project demonstrate that realizing the potential for local areas to play central roles in greenhouse gas emission reduction is limited by current policy, technology, and institutional conditions. Although this conclusion could be viewed as disappointing, the Global Change and Local Places team prefers to consider it illuminating. Under the right conditions, local actions can make substantial contributions to reducing greenhouse gas emissions. The challenge facing policy-makers at national and global levels is to develop the mechanisms that will enable and empower local action to realize its considerable potential.

REFERENCES

Intergovernmental Panel on Climate Change. 2001. *Third Assessment Report*. Vol. 1, *Climate Change 2001: The Scientific Basis*, ed. J. T. Houghton *et al.*; Vol. 2, *Climate Change 2001: Impacts, Adaptation, and Vulnerability*, ed. J. J. Mc.Carthy *et al.*; Vol. 3, *Climate Change 2001: Mitigation*, ed. B. Metz *et al.* Cambridge: Cambridge University Press.

Intergovernmental Panel on Climate Change. 2002. *Climate Change 2001: Synthesis Report,* ed. R. T. Watson *et al.* Cambridge: Cambridge University Press.

Interlaboratory Working Group. 1997. *Scenarios of U. S. Carbon Reductions: Potential Impacts of Energy-Efficient and Low-Carbon Technologies by 2010 and Beyond.* Oak Ridge and Berkeley: Oak Ridge National Laboratory and Lawrence Berkeley National Laboratory. ORNL-444 and LBNL-40533. September.

Interlaboratory Working Group. 2000. *Scenarios for a Clean Energy Future.* Oak Ridge and Berkeley: Oak Ridge National Laboratory and Lawrence Berkeley National Laboratory. ORNL/CON-476 and LBNL-44029.

Marland, G., T. A. Boden, and R. J. Andres. 2001. Global, Regional, and National Fossil Fuel CO_2 Emissions. In *Trends: A Compendium of Data on Global Change.* Oak Ridge: Oak Ridge National Laboratory Carbon Dioxide Information Analysis Center.

National Academy of Sciences 2001. 0000.

National Laboratory Directors. 1997. *Scenarios of United States Carbon Reductions: Technology Opportunities to Reduce U.S. Greenhouse Gas Emissions.* United States Department of Energy. http://www.ornl.gov/climate_change

United States Environmental Protection Agency. 2001. *Inventory of U.S. Greenhouse Gas Emissions and Sinks: 1990–1999*: Tables 1-3, 1-4. Washington: United States Environmental Protection Agency.

United States Global Change Research Program. 2000. *Climate Change Impacts on the United States: The Potential Consequences of Climate Variability and Change.* Cambridge: Cambridge University Press.

Woods & Poole Economics. 1998. *County Projections to 2025: Complete Economic and Demographic Data and Projections, 1970 to 2025, for Every County, State, and Metropolitan Area in the U. S.* Washington, D.C.: Woods & Poole Economics. http://www.woodsandpoole.com/

12

Global change and local places: lessons learned

Robert W. Kates, Thomas J. Wilbanks, and Ronald F. Abler

Four major questions have informed the Global Change and Local Places research agenda from its inception:

- How do the dynamics of greenhouse gas emissions and their driving forces differ at local scale?
- Can localities reduce their source contributions to global climate change?
- How and where does scale matter? and
- How can the capacity to study global change in localities be improved?

The project's major findings, insights, and lessons can be cast as answers to these four key questions and as three major observations regarding the ways the global and the local relate to each other, stated as variants of the familiar slogan *Think globally and act locally*.

How do the dynamics of greenhouse gas emissions and their driving forces differ at local scale?

The importance of attention to local scale lies not in uncovering differences in descriptions of greenhouse gas emissions by major categories, but in details that are often lost in larger aggregations. The details in question are often critical to designing effective mitigation strategies.

Overall, 1990 greenhouse gas emissions for the four study areas are significantly, but not greatly, different from global, national, and their respective state level emissions (Chapter 7). Local emissions differ moderately in the mix of greenhouse gases, somewhat more so in sources, and considerably more in total per capita and per square kilometer emissions. Carbon dioxide dominates the mix of greenhouse gases at our sites as it does nationally (Table 7.3). All four study areas generate carbon dioxide percentages of total local emissions within 10% of their state proportions and within 15% of the national percentage of carbon dioxide in the United States greenhouse gas mix. Greater variability exists for gases other than carbon dioxide. Methane constitutes 23% of Southwestern Kansas study area emissions, nearly twice the national component (12%), whereas for the Northwestern North Carolina and Central Pennsylvania study areas methane is 5% of each region's total emissions. Nitrous oxide makes up 3% of emissions in Northwestern Ohio and Southwestern Kansas, half again as much as the national proportion of 2% of total greenhouse gas emissions.

Emissions from fossil fuels rank high everywhere, though specific sources vary among sites, with Southwestern Kansas and Northwestern Ohio emitting smaller percentages than their respective states and the country (Table 7.4). In Southwestern Kansas agricultural sources (particularly cattle feedlots) emit twice the percentage of methane as for the entire state. The region's aquifer-based agriculture also releases water vapor, a fossil non-carbon greenhouse gas. And emissions from a coal-fired electric power plant dominate local emissions inventories despite the fact that the study area adjoins the largest natural gas field in the United States. In Northwestern North Carolina, biomass burning by local industries is a major source of carbon emissions but local forests sequester large quantities of carbon and serve as a sink for local emissions. In Northwestern Ohio, the production of lime and cement yield disproportionate percentages of greenhouse gases. At local levels, emissions of ozone-depleting chemicals have declined since 1970 though their importance will remain unclear until analysts develop methods to measure them for local areas.

Land-use and land-cover changes contribute little to emissions and to changes in reflectivity. The rates of change in land cover in the four Global Change and Local Places study areas produced only minor increases or decreases in albedo insufficient to seriously affect local radiative balances (Harrington *et al.* 1997). Nor do changes in land use or land cover appear to be major factors in explaining changes in greenhouse gas emissions through time at local scales except for intensive biomass sequestering in Northwestern North Carolina.

Per capita greenhouse gas emissions for the Southwestern Kansas and Central Pennsylvania Global Change and Local Places study areas greatly exceed national per capita estimates. The 90,000 human residents of Southwestern Kansas are outnumbered 10 to 1 by cattle, prodigious emitters of methane. The Central Pennsylvania study area exports large amounts of electrical energy generated with local coal to consumers outside its borders. Emissions per square kilometer exceed the national average in three of the four study areas, with Southwestern Kansas the exception. Within study areas, county emissions differ by as much as two orders of magnitude in per capita and in per square kilometer terms.

Despite the differences in local area emissions, could national inventories and the increasingly available state inventories be sufficient for policy purposes? The Global Change and Local Places site comparisons by greenhouse gas reveal that with the exception of methane in Southwestern Kansas, national and state 1990 estimates would suffice to inform local carbon dioxide mitigation efforts but would not be adequate for methane or for nitrous oxide. Where sources of greenhouse gases are of interest, using a national inventory to formulate policy would overlook the importance of agriculture in Southwestern Kansas, biomass burning in Northwestern North Carolina, particular production processes in Northwestern Ohio, and the overwhelming role of coal burning in Central Pennsylvania.

Moreover, national and state estimates for 1990 are quite limited in terms of identifying trends in emissions through time. For the Global Change and Local Places sites, the trends over time in local emissions based on estimates of greenhouse gas emissions for 1970, 1980, and 1990 constitute a unique data set, since most national and state estimates begin only with 1990. These trajectories differ among sites even in direction. Between 1970 and

1990, greenhouse gas emissions from Northwestern North Carolina increased more than two fold, at twice the rate for state of North Carolina and considerably faster than the increase in carbon dioxide emissions for the entire United States. Central Pennsylvania emissions also grew, but more slowly, increasing by 30% over the two decades. The Northwestern Ohio study area, on the other hand, experienced a striking decrease in greenhouse gas emissions from industrial sources from 1970 to 2000, whereas greenhouse gas emissions in the Southwestern Kansas study area decreased between 1970 and 1980 and grew again by 1990. Within all four study areas, great volatility in emissions occurred among localities as a result of changes in industrial sources. Somewhat unexpected were the large emission spikes caused by local catastrophes: chemical plant fires in Northwestern Ohio and methane releases from natural gas pipeline accidents in Southwestern Kansas.

To analyze the forces underlying these differing trends in greater detail, the local emissions inventories for three sites were transformed into activity groups (Chapter 8): (1) commercial–industrial, (2) agriculture, and (3) households, in order to identify the proximate forces that drove each activity group (Table 8.1). The mix does not change much over the two decades except in Southwestern Kansas, where notable emissions increases occurred in agriculture and households. By 1990, agriculture and households contributed 54% of the Southwestern Kansas study area emissions, compared with 39% in both Northwestern North Carolina and Northwestern Ohio.

Overall, these differences in greenhouse gas mix, sources, and trends over time are influenced by five major factors.

1. First and most important is *electricity generation and the fuels used to produce it*. All four study areas host power plants within their borders, but Northwestern North Carolina, Southwestern Kansas, and Central Pennsylvania export large amounts of electricity; as a consequence gases are emitted locally to produce power consumed elsewhere. In Northwestern Ohio, substantial use of nuclear energy to generate electricity reduces carbon emissions. The Northwestern North Carolina study area's emissions would be even larger were it not for the local use of biomass fuel in industry.

2. *Study area natural resource economies* affect emissions as well. Agriculture accounts for the distinctive emissions mix in Southwestern Kansas, and to some degree in Northwestern Ohio. Current forestry practices yield substantial carbon sequestration in Northwestern North Carolina. Local coal mining results in substantial emissions from coal combustion in the Central Pennsylvania study area.

3. *The dynamics of local economic development* also influence greenhouse gas mixes. In Southwestern Kansas, greenhouse gas emissions fluctuate with natural gas production, the growth of cattle feedlots, and the growing of the crops that feed them. In Northwestern Ohio, emission declines followed the restructuring of a regional economy traditionally dependent on such heavy industries as petroleum and steel that has now enhanced its competitiveness in automobile assembly and allied industries. In Northwestern North Carolina, much of the increase in greenhouse gas emissions is attributable to the growth of dispersed light manufacturing of furniture and food processing and to the employment they provide. In Central Pennsylvania, coal-fired energy production is the dominant driver of emissions.

4. *Changes in technology* have substantial effects on emissions. The most prominent have been changes in the fuels burned for electricity generation. From an initial heavy dependence on coal, shifts to nuclear power and to biomass burning have yielded substantial reductions in greenhouse gas emissions by electric power utilities. Other beneficial technological changes are the increased fuel efficiency of automobiles, more widespread use of natural gas for residential heating, and improved energy efficiency in manufacturing. These and similar changes have reduced the output of greenhouse gas emissions per unit of production and transportation.

5. *Growth in population* and disproportionately greater growth in the number of households has resulted in increased energy use per person. As the numbers of households have increased while the average number of people per household has declined, the amount of energy required per person has grown. Each new household requires at least a minimum of energy for basic heating, cooling, and transportation. But many households required even more energy as the average amount of space and the number of vehicles per household grew, compared with the more crowded, less mobile households of the past.

The local development, technology change, and population growth factors that strongly govern local and general trends are related to the generalized I = PAT identity that ties impacts (emissions in this case) to population growth (households in Global Change and Local Places analyses), affluence or income growth (economic development), and the degree to which technology increases or decreases emissions (Soulé and DeHart 1998). A generalized I = PAT model accounts well for the rapid growth in Northwestern North Carolina. But emissions from the Southwestern Kansas study area, with 92,000 people and 960,000 cattle, are driven more by technology and by agribusiness restructuring, and the Northwestern Ohio study area emissions are driven by changes in technology and large-scale economic restructuring. Added to these three local processes are the emissions created by producing locally materials, goods, and services for non-local markets. All four of these processes – population, consumption, technology, and export – are intermediate to ultimate driving forces that consist of a complex array of interdependent cultural, social, and economic forces (reviewed in Chapter 8).

Driving these dynamics of local economic development and associated increases in the number of households and technologies in use are five major groupings of underlying processes, many of which operate far beyond the Global Change and Local Places study areas: (1) final consumer and intermediate market demand, (2) regulation, (3) energy supply and price, (4) economic organization, and (5) social organization. Growing demand for products and services produced in the study areas is an important driver of local economic development and attendant emissions. Government regulation, from the Clean Air Act to vehicle emission standards and the National Energy Policy Act, has been a key driver of technology choices. Sometimes regulation yields perverse results, such as the construction of a coal-fired power plant on top of a large natural gas field. More often regulation has encouraged reductions in greenhouse gas emissions. Fluctuations in the price and supplies of energy have affected rates of natural gas extraction in the Southwestern Kansas study area and the use of natural gas to replace coal for home heating in Northwestern North Carolina. Under the rubric of economic organization are found such global trends as the intensified international competition in heavy manufacturing which resulted in plant closings and economic restructuring in Northwestern Ohio, as well as the emergence of the large agribusiness conglomerates that command the financial resources to expand feedlot-based animal farming and meat production in Southwestern Kansas. Finally, dramatic changes in social organization are evident in new and different employment structures and locations, in household composition, and in general norms and values.

In sum, local emissions at the four sites differ substantially but not greatly from global, national, and even state emissions. Local emissions differ moderately in the mix of greenhouse gases, somewhat more so for sources, and considerably in total emissions per capita and per square kilometer. Land-use and land-cover changes cause only minor variations in emissions and reflectivity. On the other hand, emissions trajectories from 1970 to 1990 at the four study areas differ greatly. Overall differences are due primarily to five local features: (1) whether the study area contains an electricity generation facility; (2) whether the local economy relies upon natural resource extraction and use; (3) the dynamics of local economic development; (4) changes in the numbers and sizes of local households; and (5) the technologies used over time. To some degree, electricity generation and resource exploitation are random elements, the luck of a geographical lottery that places a utility-generating plant within a study area or makes agriculture, forestry, or mining a mainstay of the local economy. Local development, households, and technology are more generic, however, being local expressions of the same trends in population, affluence, and technology found at national and international scales. These factors in turn are affected by such underlying processes as consumer and market demand, regulation, energy supply and prices, and economic and social organization.

Can localities reduce their source contributions to global climate change?

On their own, localities will find it difficult to reduce their greenhouse gas emissions over time. This conclusion emerges from four lines of evidence: (1) an assessment of future emissions growth and the specific technological and institutional potential for reducing emissions at each site (Chapter 11); (2) an exploration of the awareness and the willingness of major emitters and households to undertake emission reduction actions (Chapter 9); (3) previous experience at each site in coping with an analogous environmental problem (Chapter 10); and (4) local efforts underway to reduce emissions.

Emission abatement is primarily a local activity that necessarily takes place within the context of overall government and corporate policies. Given this reality, projecting the driving forces of local economic development, household growth, and technology change is similar to shooting at a moving target. Over the near future (1990–2020), projections suggest that greenhouse gas emissions at all study areas will increase. Projections based on econometric estimates at county scale foresee emissions increasing by about a third in all four study areas. When these estimates are supplemented by local knowledge, they yield emissions increases of 30–50% from 1990 to 2000 in three of the four study areas and a lesser rate of increase in Southwestern Kansas.

Adopting as illustrative goals the 7% reduction from 1990 emissions by 2010 specified in the Kyoto agreement for the United States, or the more challenging 20% reduction by 2020 agreed to by many cities, will require overall reductions from 1990 emission levels of 13–37% by 2020 to reach a 7% reduction target under the local knowledge emission projection. An overall reduction in emissions of 25–46% will be required to attain the 20% target. How difficult would it be to achieve such reductions, what role does local knowledge play in the process, and how willing are study area residents to make such efforts?

Few of the prerequisite incentives and none of the mandates that might support local action are currently in place in the four Global Change and Local Places study areas, except for Toledo Ohio's membership in the Cities for Climate Protection program and the role of BPAmoco in the Northwestern Ohio study area. Examples of incentives that could be effective are tax credits for emission-reducing investments and charges for carbon emissions. Examples of mandates include compulsory emission limitations and more stringent motor vehicle emission standards. One example of a combination of mandates and incentives is an emission cap, defined in terms of permits to emit that allows permits to be traded. Another example, embodied in the philosophy of the Clean Development Mechanism (CDM), would permit United States emitters to reduce emissions either from their own operations or from other emitting sources (possibly including biomass sequestration) elsewhere in the world.

Several study areas offer technological possibilities for addressing major sources of greenhouse gases. Examples include the substitution of natural gas for coal in power plants in Northwestern North Carolina and Southwestern Kansas, industrial efficiency improvements in Central Pennsylvania and Northwestern Ohio, transportation efficiency improvements in Northwestern North Carolina, Northwestern Ohio, and Southwestern Kansas, and reductions of methane emissions from cattle in Southwestern Kansas. In at least two study areas, applying known mitigation abatement strategies to the site-specific mix of gases and sources might reduce building energy use emissions by as much as 20–50% and transportation energy use emissions by as much as 50% by 2020 according to analyses based on the standard Cities for Climate Protection methodology, although the Global Change and Local Places study area teams doubt that these goals would be realized. Overall, it is possible that a scale bias exists in estimating potentials from local-scale initiatives. Local potentials may be overestimated by global or national analysts who fail to appreciate local complexities, while potentials may be underestimated by local analysts whose immersion in area complexities may lead them to discount changes in external driving forces.

Site-specific mitigation opportunities can further reduce emissions in some study areas. Identifying site-specific sources and opportunities is a substantial step toward assessing local capability and inclination to help abate greenhouse gas contributions. Focusing attention on such place-specific concerns as the net carbon yields from biomass burning in Northwestern North Carolina (Lineback *et al.* 2000), the efficiency of natural gas pipeline compressors in Southwestern Kansas, the shifts in industrial mix and location in Northwestern Ohio, or the potential negative local impacts of national mitigation strategies in Central Pennsylvania directs the search for realistic and effective local options for mitigation. Understanding the complex driving forces of greenhouse gas emissions broadens the search even further – suggesting how opportunities for growth management in Northwestern North Carolina, irrigation efficiency in Southwestern Kansas, sprawl limitation in Northwestern Ohio, and economic diversification in Central Pennsylvania can lead to substantial emission abatement while at the same time addressing the larger problems for which such policies are proposed.

Yet the decisions to utilize these strategies may be diffuse, with local emissions occurring far from the loci of their control. Greater or lesser shares of local emissions may originate in production that serves non-local markets: natural gas production in Southwestern Kansas, automobile assembly in Northwestern Ohio, and furniture production in Northwestern North Carolina. Similarly, local consumption causes emissions elsewhere, as in the instance of local use of electricity generated outside a study area. Small non-local sources contribute to emissions, such as vehicles traversing an area. Even more difficult to identify are the locations where decisions to engage in emission-producing activities are made, and where decisions to reduce emissions might be made. Findings to date suggest that in general, localities have limited control over the forces driving local emissions (Chapter 9).

Focusing on the greenhouse gas emissions that *are* partly or wholly controlled locally shifts attention to how willing local entrepreneurs and residents are to reduce emissions. To explore local willingness to undertake emissions abatement, three specific lines of inquiry were pursued in each study area: (1) surveys of major emitters and households, (2) case studies of responses to analogous environmental concerns, and (3) current participation in local efforts to reduce emissions.

Evaluating potentials for local action is especially complicated for an emerging issue such as reducing greenhouse gas emissions. When an environmental issue first becomes salient, as contrasted with a relatively mature issue familiar to most stakeholders and well supported by information and evidence, assessing local attitudes toward doing something about the issue on the basis of a single survey is of limited value. Such a survey cannot be expected reliably to predict attitudes toward the issue as it matures and can, therefore, provide little insight into potentials for local action at later dates. Surveys on mature issues familiar to respondents yield more reliable conclusions. In our judgment, climate change issues are still emerging.

The major emitters surveyed by Global Change and Local Places analysts included such major energy providers as electric power generators, natural gas producers, and a large university; the managers of manufacturing and oil companies' industry facilities, transportation (trucking) company officials, and feedlot operators. Face-to-face or telephone interviews with local mangers or representatives explored trends within their industries, their awareness of global warming and degree of concern about its effects, the locus of control over emission generating and mitigating decisions, and their views on specific opportunities for mitigation. Levels of concern among major emitters ranged from considerable skepticism about global warming in Southwestern Kansas to informed efforts at reducing emissions in Northwestern Ohio. In general, the largest industry emitters were well informed about climate change issues and potential regulation resulting from the Kyoto accord. Smaller emitters focused more on how air pollution concerns would affect their operations than on the possible effects of climate change. Even the most informed, however, saw little direct connection between global warming and their local activities. While fuel and technology switching in electrical generation offer similar mitigation opportunities in all study areas, the potential for other effective actions (most often some form of energy efficiency improvement) differs considerably among study areas. Although the decision authority for many of these technological opportunities rested beyond the local managers, they expressed strong preferences for state and local regulation rather than federal oversight, deeming state and local officials as more trustworthy and more likely to understand local problems and capabilities.

In the Global Change and Local Places household surveys, global warming and human-induced climate change ranked low among householder concerns, indeed much lower than in national opinion polls, which reveal higher levels of concern and greater willingness to reduce emissions, even at greater individual cost. Mitigation opportunities for households were similar in all four study areas, involving more efficient energy use for household transportation, residential space heating, and electricity use. There was little willingness among householders to adopt such abatement strategies, however, because of low levels of concern about climate change.

Another approach for gauging local propensities to mitigate emissions used historic analogs of local responses to major environmental problems (Chapter 10): the drawdown of the Ogallala aquifer irrigation water in Southwestern Kansas, watershed protection in Northwestern North Carolina, Lake Erie pollution in Northwestern Ohio, and air pollution abatement in Central Pennsylvania. These analog studies sought to identify the elements of success in the historic cases and what they might suggest regarding capacities and propensities for local responses to global warming.

The Ogallala aquifer contains fossil water trapped millions of years ago. It underlies much of the Southwestern Kansas study area. Since the Ogallala absorbs minimal recharge from rainfall, the region is essentially mining its groundwater. As an analog to global warming, groundwater depletion is a gradual, continuous, long-term, and cumulative process that is difficult to reverse and that threatens the long-term sustainability of agriculture in the Southwestern Kansas study area. In response to dropping water tables as well as to the high energy costs incurred in pumping ground water and low crop prices, local irrigators have adopted water conservation practices that have reduced groundwater use. In one of the Southwestern Kansas counties, the drawdown of the aquifer has declined from almost 0.5 meters per year to 0.08 meters per year over a ten-year period. The Southwestern Kansas analog study suggests that if irrigators were to accept global warming and its potential negative impacts on agriculture as a reality, they would be receptive to mitigation and adaptation activities that work to their economic advantage, much as they have done in response to Ogallala ground water depletion.

In Northwestern North Carolina, efforts to safeguard the quality of water supplies by protecting their watersheds provide an analog to greenhouse gas mitigation. Growing concerns about quality and quantity of water supplies in the rapidly growing eastern part of the state served as the catalyst for what ultimately became the landmark 1989 Water Supply Watershed Protection Act. It took seven years from the first statewide expressions of concern over water supply quality to achieve passage of the Act, and until 1994 for full enactment of the Act's restrictions. Despite the eleven-year genesis of the program, its achievement offers encouragement regarding responses to climate change: once they became aware of the problem, communities and the state acted to abate declining water quality and quantity in advance of federal regulation and later found ways to accommodate federal mandates.

The Northwestern Ohio analog focuses on the cleanup of Lake Erie. The Maumee River in the Global Change and Local Places study area was a major contributor of suspended sediments, phosphorus, and toxic chemical loads that by the 1970s had turned Lake Erie into a dead lake. Phosphorus-fertilized algae blooms consumed the Lake's dissolved oxygen when they decayed, which in turn led to the loss of numerous desirable fish species. Reducing pollutants by 10% took a decade, and the overall cleanup of the lake took thirty years from widespread recognition of the problem to the improved beaches, clearer water, good fishing, and supportive wildlife habitat Lake Erie now offers. The cleanup required both global and local action: a Canada – United States effort that defined the general policy goals for the Great Lakes, and local action committees composed of key stakeholders that undertook the specific programs needed to meet those goals.

The Central Pennsylvania analog is the highly localized impacts that would follow nationally mandated mitigation regulations. The coals mined in the Central Pennsylvania study area contain relatively high percentages of sulfur. Although coal consumption continues to increase in the United States, Pennsylvania coal production and employment have decreased substantially in part owing to the Clean Air Act and its restrictions on sulfur dioxide emissions. Mandated reductions in coal use – one way to mitigate greenhouse gas emissions – could trigger further declines in coal production in the parts of the Central Pennsylvania study area that already suffer high unemployment and low average incomes. Without proactive technological improvements to ensure continuation of the coal industry or the provision of a strong, fine-meshed socioeconomic safety net, already declining counties and communities could face ruinous economic decline.

Beyond these analogs, two study area teams became involved in real time efforts at mitigation. In Northwestern Ohio, the City of Toledo has joined the International Council on Local Environmental Initiatives Cities for Climate Protection Program. The Global Change and Local Places Northwestern Ohio study area team undertook an extensive and collaborative set of interviews with the major emitters of the area. In North Carolina, the Global Change and Local Places study area team, with United States Environmental Protection Agency funding, has prepared an *Action Plan for Reducing North Carolina's Greenhouse Gas Emissions*. At another site, Southwestern Kansas research findings encouraged action by members of the state's congressional delegation. These examples suggest that researchers from the local area in collaboration with former students and local decision makers can elicit information from area residents that would not be forthcoming to unknown, non-local experts. In both these instances, local researchers became respected sources of technical information and policy advice.

In sum, significant emission abatement will be difficult for local places to accomplish if they are left to their own devices. As outlined in Chapter 11, local actions to reduce greenhouse gas emissions could yield substantial results under four sets of conditions: (1) growing evidence of impacts, related to climate change forecasts and impact assessments; (2) policy interventions that directly or indirectly associate emission reductions with local benefits;

(3) incentives and assistance for local innovation; and (4) technology improvements. But continued growth means that meeting even the modest targets specified in the Kyoto or Toronto agreements will require access to technological and institutional resources that currently lie beyond local reach. Localities lack control over significant proportions of their emissions, and they currently do not exhibit the understanding of global climate change and the will to abate and remediate emissions that are prerequisite to effective action, even if they acquired local control over emissions. The portfolio of available technological opportunities is a poor match for the abatement needs of the Global Change and Local Places study areas in most respects, and the needed institutional framework of incentives and mandates is still absent. Most realms in which political or economic decisions are made do not match well the county boundaries of the study areas, and many of the major decisions needed to abate emissions will be made outside the study areas boundaries if they are made at all. As of 2000, resident knowledge of global warming and concern about its effects was limited. Even well-informed major emitters were usually reluctant or unable to undertake abatement measures on their own. However the analog studies do offer the encouragement of demonstrating that a combination of global and local actions can yield positive results when the local impacts of environmental change become evident. Similarly encouraging are the efforts undertaken by local corporate affiliates in Northwestern Ohio in response to policies adopted at the corporate level.

How and where does scale matter?

In light of what the Global Change and Local Places Project has revealed, what of the grand query that gave birth to it (Chapter 1)? In theory, scale matters in studying global change, local dynamics are worthy of concern, and localities can make a difference. It is clear that some of the forces driving global change operate at global scale, as exemplified by the greenhouse gas composition of the atmosphere and the reach of global financial systems. It seems equally clear that many of the individual phenomena that underlie micro-scale environmental processes, economic activities, resource use, and population dynamics arise at a local scale.

The literatures of global change, climate change, natural hazards, and environmental studies describe two kinds of ways scale seems to matter in global climate change (Wilbanks and Kates 1999). First, climate change results from the interaction of the two different domains of nature and society, each composed of many systems operating at different scales of space and time that result often in an inherent mismatch in scale between cause and consequence. Second, in the current practice of science, studies relating global climate change to local places have been conducted top-down, moving from global toward local scales, using climate change scenarios derived from global models as starting points, despite the absence of regional or local specificity in such models. At global scale, understanding the complexities among the environmental, economic, and social processes that induce change seems intractable. Thus more focused, place-based research seems to make tracing these complex relationships more feasible, especially when the researchers are armed with specific local knowledge.

From the bottom up perspective, although isolated case studies may be more readily understood than global integrated analyses, they are also much less generalizable. The generalizability of such studies is enhanced where they are viewed as natural experiments, which have been carefully chosen for comparability, use a common study protocol and are compared with control studies. Local areas also offer a greater variety of information and experience than that found in a sample of larger areas. This relationship provides an opportunity for detailed understanding of the complex causes and consequences of global climate change, understanding that often remains concealed in processes or relationships averaged over larger spaces. Locality studies also offer different perspectives; researchers examining an issue from a local perspective often come to conclusions different from those reached by analysts who view the same issue from a global vantage point.

The causal chain associated with global climate change includes six links: (1) driving forces; (2) human activities; (3) radiative forcing; (4) climate change; (5) climate change impacts; (6) and responses (Figure 1.2, p. 6). The Global Change and Local Places research team examined the geographic scale of actions for each link and asked how well current scales of assessment – observation, research, and policy – match the scales of each of the six processes. As shown in Figure 1.2, the geographic scale of actions varies considerably between links, and an envelope of the large-scale actions is a wave in which global and large region actions fit the driving forces of population, affluence, and technological change, the radiative forcing of gases, aerosols, and reflectivity, climate change, and preventive and adaptive responses. In contrast, emissions and sinks and the major impacts of climate change are far more localized. Given the general bias toward the global and large region of current scales of assessment, serious scale mismatches exist between activities and assessments, particularly for impacts and specific responses other than enabling or requiring actions. Local scale, for instance, seems to be especially important at the upstream (driving forces, human activities, emission generation) end and the downstream (mitigation and adaptation responses) end of the cause–consequence continuum. The problems that such scale disjunctures cause are now being addressed by efforts underway to move downscale in each of the causal domains, while studies such as the Global Change and Local Places project seek to move upscale.

From this and related studies, our findings suggest that local scale matters greatly in several major ways. Local scale provides a test bed in which to pose alternative proposals for emission abatement and for adaptation that can serve to bound what is realistic and what is possible in emission mitigation and adaptation. Local scale can permit identification of new or understudied opportunities for mitigation and adaptation, as shown by the development of a niche industry that provides waste biomass as a fossil fuel substitute in Northwestern North Carolina. Most important, the research process itself transmits information and understanding, often from a trusted source, thus making global–local research a contribution to the social process of informing the citizenry and – where the information is adequate – creating essential support for action at state and national scales.

Local studies can also reveal the differential local consequences of national and state actions. The existence of a coal-fired plant in the heart of a large natural gas field in

Southwestern Kansas can be traced to well-intentioned but misguided efforts to address the energy crisis of the 1970s. Needed care in the development of regulations is suggested by the inequity of the local burden that a national carbon permit or tax system would place on the economy of Central Pennsylvania.

In two important dimensions of understanding global change, Global Change and Local Places project experience suggests that local studies are *not* essential: (1) greenhouse gas emissions can be estimated on a national or state basis and then further aggregated globally for purposes of modeling climate change, understanding the relative contributions to emissions from large regions, and for negotiating international agreements; and (2) local studies are not prerequisite to understanding the underlying processes that structure the proximate driving forces of local economic development, household growth, and technology choice. Locality studies are fruitful, but not for every purpose.

Emission reduction, however, is almost always deeply imbedded in such cross-scale interactions as the intertwining of local activity with larger-scale government and corporate policies. The development of a technology portfolio for mitigation and adaptation, for example, relies primarily on initiatives extending beyond localities, on such factors as basic research, government programs and incentives, and corporate research and development. Local knowledge can be helpful in suggesting under-recognized needs and by offering local innovations for further development, but it cannot itself respond to such needs or disseminate such innovations.

The Global Change and Local Places experience demonstrates that different scales are needed in examining different functions, the key being the capacity of a local area unit to act upon emission generation, mitigation, or climate change adaptation. Such functions as setting abatement goals, regulations, incentives and negotiating international agreements are clearly best done at national scale, although similar actions are also undertaken within large transnational corporations. States and large regions have considerable decision-making capacity. They are the main regulators of land use, but can also strongly influence emissions, technology choices, and the like by regulating means to achieve air pollution standards or specifying renewables in electricity mixes. Regions (often aggregates of states) seem appropriate for examining likely impacts and developing some adaptation strategies. Within states, congressional districts are crucial for the political discourse needed to support efforts to abate emissions or adapt to their consequences. For specific emission-generating activities, such distinctive geographical units as electricity sheds or grids, pipelines, sewage disposal, and transportation districts may be relevant. Counties, especially where data on fuel consumption are available, are basic building blocks for calculating emissions. Depending on where they are located, counties may possess some direct decision-making capability. Cities, as exemplified in the International Council for Local Environmental Initiatives Cities for Climate Protection Program, may exhibit surprising abilities to encourage mitigation and adaptation. Finally, depending on the action desired or required, individuals, households, and firms hold the ultimate decision-making power in much mitigation and adaptation.

Simply stated, the Global Change and Local Places experience suggests that cross-scale interactions are more significant than aggregate differences among scales in understanding global climate responses. If that inference is accurate, integrating global change analyses in terms of a single scale metric may be highly questionable. Integration conducted at a single scale will result in the loss of essential information about cross-scale interactions.

The 1° scale adopted for the Global Change and Local Places study areas was an arbitrary starting point for investigating scale issues. That scale was selected in part to assure diversity within each study area and to link to potential global change model grid sizes. In the course of the project, other local scales were examined as well. Wide differences among counties were found in all study areas and among cities in the Cities for Climate Protection campaign. The Central Pennsylvania study area was also part of a large complementary watershed study. At the outset (February 1997), and again in 1998, Global Change and Local Places project leaders brought together several groups that had also adopted bottom-up, locality-based approaches. These included Clark University's Critical Zones Project, the International Geosphere–Biosphere Project on Land Use and Cover Conversion (LUCC) project, The International Council for Local Environmental Initiatives Cities for Climate Protection Project; the Global Assessment and Local Response Project of Harvard's Kennedy School of Government, various Canadian regional impact studies, and European projects related to the broader issues of sustainability. In March 1998, groups from Clark University, Harvard University, and the Global Change and Local Places study areas met in Boston to consider impact assessment issues from the perspective of local places. Each of these efforts focused on different scales. At the same time, top-down international and national assessments were initiating major efforts at downscaling (Chapter 1). Thus, concurrent with the Global Change and Local Places analyses, larger regional efforts were underway, including the United States Environmental Protection Agency sponsored emission studies and action plans in many states, as well as the nineteen geographic region studies that were part of the United States National Assessment of Climate Change Impacts (United States Global Change Research Program 2000). Most recently, in August 2001, the governors of the New England states joined with the premiers of Eastern Canada's provinces to adopt a climate action plan to reduce regional greenhouse gasses by 2020 to a level 10% lower than that of 1990.

In summary, scale matters. It matters across science and policy, in theory and in practice. Local scale especially matters at the upstream (emission generation) end and the downstream (mitigation and adaptation) end of the cause–consequence continuum. Local studies make it possible to bound what is realistic and what is possible in emission mitigation and adaptation, to identify new or understudied opportunities for mitigation and adaptation, to transmit information and understanding to local officials and residents, and to create support for national and state policies. Local studies can be helpful in setting regulations and incentives by identifying the different local consequences of national and state policies and programs. Local studies can help in a limited way in the creation of a technology portfolio for mitigation and adaptation by suggesting under-recognized needs and local innovations. Local studies are not important for estimating greenhouse gas emissions on a national or state basis or for understanding major economic and demographic driving forces. There is no ideal local scale, but different scales are needed for different functions, the key being the capacity of the local unit to act upon emission generation, mitigation, or climate change adaptation.

How can the capacity to study global change and local places be improved?

Global Change and Local Places's original vision included learning how to improve the capacity to study global change in local places. Essentially, the AAG project team sought to use its project experience as a basis for improving what is known about research strategies, institutional contexts, and research protocols that might be applied to local case studies (Chapter 2). Some aspects of that vision were realized, others were not, some major elements remain elusive, and some efforts needed to augment the capacity to do local studies are already underway.

Most gratifying was the demonstration that the original expectation that local knowledge would inform global change studies at all scales and that institutions with continuing commitments to local area studies and outreach would be invaluable in uncovering such local knowledge was sound. Long-term global change studies at local scale must be hosted in institutions with continuing commitments to their regions if those studies are to be sustained as individual key investigators come and go. Moreover, based on Global Change and Local Places experience, local research on global change is heavily dependent on existing local information infrastructures. In the Global Change and Local Places study areas, investigations could not have been conducted effectively if they were limited to statistical data available at the county level from national and state sources. Local informants were essential for framing the issues, understanding local contexts, uncovering inventories and other data sources available only at a local or small-regional scales, and interpreting the local meaning of numbers available from more general sources, especially over periods of two decades or more. The personal knowledge and expertise brought to Global Change and Local Places analyses by local team members was almost always indispensable to identifying and tapping local information infrastructures, formal and informal.

The Global Change and Local Places experience also suggests that, consistent with other findings about the assessment process, eliciting local knowledge is facilitated by using local experts as gatekeepers. Recent research in studies of environmental assessments worldwide reveals that the results of assessments are much more likely to be put to use in local areas if they are channeled through local experts. The same principle applies in the opposite direction if Global Change and Local Places results are any guide. Local experts are uniquely suited to assist in accessing local knowledge, because they are repositories of so much of that knowledge and because their local contact networks – usually well salted with former students from local institutions – include the salient local information infrastructures.

Involving institutions with a commitment to their local areas works, confirming one of the initial Global Change and Local Places premises. An effective way to get local experts involved in bottom-up assessments of global change issues is to bring local universities into the process as major participants. In some cases, these universities may focus primarily on their own regions. In others, the universities may be leaders in global environmental research and development while still maintaining considerable interest in their own regions in connection with land grant extension missions. What these institutions have in common is a commitment to long-term local and regional well-being, to staying in touch with local concerns, to bringing knowledge to bear for local improvements, and to continuing relevant programs of teaching and learning over the long term as individual leaders and experts change through time. In many cases, of course, they are public institutions. Project results make it clear that faculty in institutions that have historically been region-oriented and connected only intermittently to national and global research enterprises are as capable of producing first-class research as colleagues who are more widely known in the research community, suggesting in turn that differences between prominent and less well-known researchers are partly a matter of differences in opportunity.

If the Global Change and Local Places team were redoing the study, we would neither choose 1° grid study areas nor would we attempt to downscale the existing Intergovernmental Panel on Climate Change/Environmental Protection Agency emissions methodologies. The initial selection of 1° equatorial latitude–longitude study areas was an arbitrary but well-intentioned choice made to permit Global Change and Local Places study areas to be embedded into the grids of climate models and to enable Global Change and Local Places findings to inform efforts to downscale models beyond their then current 5–10° grids. In practice, there is no ideal scale for studying global change in local places, and the arbitrary 1° scale has no intrinsic value. Valid arguments can be made for either larger or smaller scales, or for boundary modifications to include or exclude activities of interest that have particular weight and might therefore seriously affect general findings. Regardless of study area sizes, boundary issues are salient at local scales because external connections become increasingly important as smaller and smaller areas are examined. Accidents of boundary determination that include or exclude such individual point sources of greenhouse gas emissions as coal-fired electric power plants, for instance, powerfully effect locality emission totals. Somewhat larger areas would in some cases reduce disparities between emission production and end-use emission figures, while smaller areas would facilitate focus on particular issues or systems, such as agricultural processing in Southwestern Kansas.

A serious problem with using a latitude–longitude oriented scale for local studies, regardless of size, is that the scale is unlikely to approximate the size and boundaries of any significant geographic unit for which or in which decisions are made. For example, the Cities for Climate Protection program focuses on incorporated cities, counties, and metropolitan areas – units for which actions based upon the results of a study are possible. As a general rule, the Global Change and Local Places experience suggests that if the intent of a study is to inform decision-making, relating the scale of study areas to the scale of decision-making units offers great benefits. In the United States, some studies of global climate change impacts might best be conducted at the scale of Congressional districts if the purpose is to inform national policy-making.

The Global Change and Local Places analyses also revealed that current Intergovernmental Panel on Climate Change/Environmental Protection Agency protocols for state emissions inventories are too complicated and data-intensive for most locality studies. Even in the data-rich United States, many critical data sets are not available at local or county levels. Examples abound: fossil fuel consumption – the major single variable in emissions – is not tabulated locally; electricity generation and consumption data rarely fit the scale of local study areas; and ozone-depleting chemical emission estimates rely on national-scale production data (Kates *et al.* 1998). Even when emissions data can be allocated locally, they are not available in ways of interest to the people and the organizations that produce emissions and can potentially control and mitigate them. For example, householders cannot be told for which emissions they are responsible. Only fossil fuel consumed in the home is classified as residential. Emissions from household electricity consumption are estimated within a category of utilities; energy consumption for household transportation is estimated within a general transportation category; some methane originating in households is measured as waste in municipal landfills; and household ozone-depleting chemicals are estimated from per capita allocations of national production, much of which is actually consumed industrially. To remedy these and other difficulties, Global Change and Local Places team members devoted much time and effort to creating a downscaled version of the Intergovernmental Panel on Climate Change/Environmental Protection Agency emissions methodology suitable for use at the scale of a United States county. Were the Global Change and Local Places team to redo the analysis, it would begin with a bottom-up emissions inventory, a variant of which is used by many cities that participate in the International Council for Local Environmental Initiatives program.

Most important, were we to redo the project, we would place major emphasis on local impacts of climate change, vulnerability to such climate change, and adaptive as well as preventive or mitigative responses. When Global Change and Local Places was designed in 1995–1996, observations of the impacts of climate change on natural and human systems for its prospective study areas had not been made. To simulate what might occur seemed to require reasonably reliable climate change forecasts or compelling analogs of possible climate change. Such forecasts were not available at the scale of the Global Change and Local Places study areas at that time, nor is there yet a common analog that is valid beyond particular regions. Thus a climate change impact dimension was not included in this study, although provision was made to do so in any future follow-up studies that may be undertaken. If we were to design the Global Change and Local Places project today, we would employ a modified concept (Figure 12.1) that would include climate change impacts, vulnerabilities, and the considerable potential for surprise, instead of the original schema (Figure 2.9, p. 36).

Since 1995–1996, three important changes in research and observation have taken place which today would easily support a major impacts dimension to a study similar to Global Change and Local Places: (1) impacts on natural systems, while still not widespread, are increasingly being observed in the form of longer frost-free periods, early thaws of frozen rivers, and the migrations of animals, birds, insects, and plants; (2) extreme climate events have been widely experienced and these might serve as local analogs for increased climate variability; and (3) beyond observation, climate change forecasts are becoming available at scales of $0.5°$ or even smaller. Even more important has been the widespread acceptance of a vulnerability approach to local areas as a way of bridging the gap between actual observations of impacts and adaptive responses to them, and the simulation of impacts that require local forecasts from global climate models. Vulnerability analysis of a study area begins with examination of the relative sensitivity to climate change of the particular

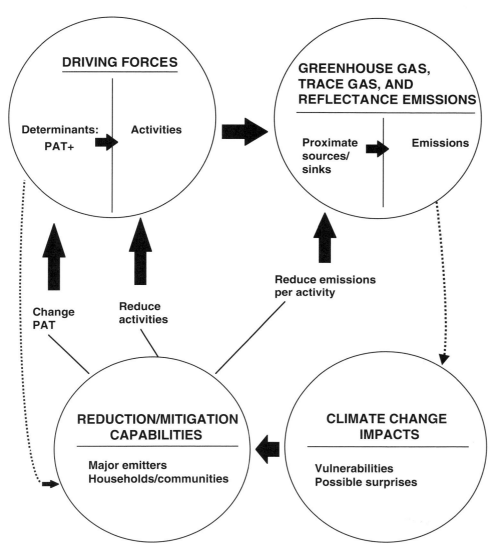

Figure 12.1 A modified Global Change and Local Places concept.

ecosystems, populations, and activities found therein and an assessment of their capacity to cope with climate change or to respond to it. In the recently completed United States National Assessment of Climate Change, local knowledge on the part of stakeholders was found to be extremely useful in identifying regional vulnerabilities and sensitivities to climate change, and in assessing the capacity to cope with possible impacts. Over the next several years, vulnerability analysis methodology will become much improved (Kasperson and Kasperson 2001) and should be used for future local study areas. These changes in impacts and vulnerability analysis are part of a growing capacity to undertake local area studies that has taken three promising directions: (1) the creation of new tools and methods,

(2) expanding collaborative networks, and (3) new sources of support to fund and encourage local studies.

Local areas can take many forms ranging from individual households to congressional or parliamentary districts, each of which requires methods and tools appropriate to its scale, functions, and comparative advantages. The necessary suite of methods and tools is expanding, but it is far from complete, especially if needed impact and vulnerability analysis is included. For emissions from local areas such as counties and groups of counties, the Global Change and Local Places downscaling of the United States Environmental Protection Agency's state version of the international (Intergovernmental Panel on Climate Change) guidelines is generally workable in the United States assuming at least a moderate level of financial support, although some problems remain to be resolved. For example, both production- and consumption-related emission inventories were developed, but realistic ways of allocating ozone-depleting chemicals to local places are still needed. The single action that would most improve local data would be to extend reporting requirements for state energy consumption data down to the county level.

A better alternative for emissions is to use the International Council for Local Environmental Initiatives Cities for Climate Protection emissions inventory, which employs simplified software developed by Torrie-Smith Associates for estimating carbon dioxide emissions and methane emissions from energy production and use, transportation, buildings, and households and for identifying locally implementable measures to reduce those emissions (Kates *et al.* 1998). The International Council for Local Environmental Initiatives software has now been used in 504 cities and metropolitan areas around the world (115 in the United States). Originally devised for cities, it has recently been expanded for use in a statewide study of greenhouse gas reduction strategies in North Carolina (Appalachian State University Department of Geography and Planning 2000) and now includes emissions from agriculture and utilities. A growing number of corporate emissions protocols are available (Loretti *et al.* 2001; Margolick and Russell 2001) as well as calculators for individuals and households. Beyond emission protocols, the current availability of methods and tools is less promising. The International Council for Local Environmental Initiatives package includes emission reduction measures, but generally, there is no methodology for local area impact studies or for the assessment of vulnerability.

But even when better methods and tools become available, the heart of a local area study is not accessing and arraying numerical data from readily available tables according to a standardized methodology. Rather, the goal of such efforts is a series of judgment decisions about priorities in terms of the major questions being asked, the effort to be devoted to filling gaps in available data, and the importance of precision and comparability in particular instances. The Global Change and Local Places project addressed the issue of how to improve prospects that local area studies of global change, conducted through the judgments of different people in different study areas, can produce results that are sufficiently comparable to permit the development of broader understandings. If those ends are to be achieved, the Global Change and Local Places premise was that the individual studies need to ask similar questions, generate data in similar categories based on similar techniques for measurement or estimation, and make data available in similar formats. To achieve that kind of comparability, a protocol is needed that describes a process for conducting a local area study related to global changes. Such a process protocol would guide a local area study team through the steps to be taken and the questions to be considered, and specify the judgment decisions to be expected and guidelines for making them. In effect, such a protocol would be more of an expert systems shell than a set of footnotes for a matrix of numbers. The Global Change and Local Places team hoped that such a protocol would result from its studies. Though we have learned much about creating such a protocol, we have not actually done so. The task turned out to be far more complex than initially expected, and its requirements exceeded available concepts, tools, and methods.

Beyond improved concepts, methods, tools, and protocols, the capacity to undertake local area studies is being expanded through a variety of promising collaborative networks and programs that encourage and fund local study. In the United States, a direct outgrowth of the Global Change and Local Places project is the Human Environment Regional Observatory (HERO), a collaborative effort of The University of Arizona, Clark University, Kansas State University, and Penn State University designed to develop infrastructure for studying the long-term implications of human dimensions of global environmental change at small regional and local scales (http://hero.geog.psu.edu/). A somewhat parallel effort has been the extension of the National Science Foundation Long-Term Ecological Research Stations to urban areas (initially Baltimore and Phoenix) with much greater attention to interactions with the human environment (http://lternet.edu/). In the wake of the National Assessment of the Potential Consequences of Climate Variability and Change (http://www.usgcrp.gov/usgcrp/nacc/default.htm), new collaborations have developed among the nineteen study regions, bringing together university researchers, government officials, and various stakeholders.

Two collaborative programs are emerging that seek to augment assessment capacities in developing countries. One, the Assessment of Impacts of and Adaptation to Climate Change (AIACC), is an outgrowth of the International Panel on Climate Change that is funded by the Global Environment Facility (GEF) specifically to build capacity in developing countries to assess climate change impacts and adaptation potentials. The AIACC project is being implemented by the United Nations Environment Program and co-executed by the SysTem for Analysis, Research, and Training (START) and the Third World Academy of Sciences. The third, the Millennium Ecosystem Assessment, a major four-year international assessment of the state of the world's ecosystems, includes a significant sub-global assessment component that will include regional and local studies in Southeast Asia, South Africa, and perhaps other regions (http://www.millenniumassessment.org). Local capacity building is a major aim of the Millennium Ecosystem Assessment. Clearly, the development of improved capacities to undertake local area studies in developing countries poses major challenges for the future.

In sum, much of Global Change and Local Places leadership's original vision of improving capacities to undertake local area studies has been borne out by this study. Local knowledge enriches the understanding of global changes and informs discussions of what to do about those changes. Local institutions that know their areas are indeed fonts of local knowledge. Local experts serve as credible sources of information for local residents. The size the Global Change and Local Places team chose for local areas was not optimal, nor was downscaling existing emission protocols the most efficient way to compile study area emission inventories. The project would have benefited greatly from inclusion of a module that focused on impacts and vulnerability. These shortcomings notwithstanding, the Global Change and Local Places study has provided an initial set of concepts, methods, tools, a procedure for undertaking local area studies, and identified useful directions for extending these results. Moreover, it has provided an early example of fruitful collaborative workshops that have been emulated elsewhere.

Lessons learned

Simply stated, the Global Change and Local Places project shows that local knowledge is important, although not for everything. The beguiling slogan *think globally and act locally* does not suffice for climate change, its causes, and its consequences, because global or even national knowledge averages too many distinctive local trajectories of greenhouse gas emission and their driving forces, overlooking positive mitigation opportunities and making local actions more difficult. Local knowledge, however, is also an incomplete basis for action, since most decisions that could result in major local emission reductions will not be made locally. Revising the familiar slogan to recognize how the local relates to the global calls for three insights:

1. Make the global local

To think globally in a local place, the global must be made local, preferably by a local and trusted source such as the faculty members in the Global Change and Local Places regional institutions. In the course of local research, current scientific understandings are transmitted, local responsibility for emissions is outlined, and local sectors and places vulnerable to the consequences of global climate change are identified. Local studies transmit information and understanding, create support for national and state abatement and mitigation policies, and encourage local action. Local studies can identify the realistic possible emission mitigation and climate adaptation options, assess local awareness and propensities to undertake action, and explore the variable and sometimes unexpected local consequences of national and state policies and regulations. Local studies can find new or understudied opportunities for mitigation and adaptation, suggest the technology portfolio needed to take advantage of such opportunities, and even, on occasion, stimulate local innovations to address them.

2. Look beyond the local

People in localities must also think globally. Understanding distinctive local trajectories of greenhouse gases over time requires an appreciation of driving forces that operate at scales beyond the local. The dynamics of local economic development and associated increases in the number of households, for example, vary widely among study areas, as do the technologies in use. These forces in turn are driven by larger regional, national, and global processes related to consumer and market demand for products that embody emission releases, energy, environmental, and land-use regulations, energy supply and prices, industrial and economic organization, and social changes in work, household structure, and associated norms and values.

3. Act globally in order to act locally

While emission abatement or mitigation are primarily local activities, they take place within a structure and context of larger regional, national, and global government and corporate policies and programs. For localities to take steps to reduce their greenhouse gas emissions, they must develop the desire to do so, gain some control over a significant portion of their emissions, and, most important, have access to the technological and institutional means of doing so. This requires global acts by corporations, states, national governments, and international organizations that enable and encourage localities to take action. Currently, an institutional framework of incentives and mandates that would encourage local action is largely absent, and the portfolio of available technological opportunities is a poor match for the abatement needs and potentials of many local places. Despite the limited progress toward creating the necessary institutions and technologies, examples of new combinations of global and local linkages abound. We write this final paragraph in the hot summer of 2001 in the United States – the nation unique among the signatories of the Framework Convention on Climate Change in not seeking ratification of the Kyoto Protocol to reduce greenhouse gas emissions. At the same time, cities, corporations, congregations, and groups of regional leaders, enabled or encouraged by their global and national counterparts, are moving to reduce greenhouse gases in excess of the Kyoto targets; the conference of New England Governors and eastern Canada Premiers have pledged to reduce regional greenhouse gases 10% from 1990 levels. Sixty-eight United States cities and counties that are home to 28.5 million people are participating in the International Council for Local Environmental Initiatives Cities for Climate Protection Campaign and have already reduced their emissions by an estimated 7.5 million tons of carbon equivalent. Among corporations, BPAmoco Oil is currently conducting a major intra-corporation emissions cap and trade program that aims to reduce the corporation's emissions by 10% by 2010. BPAmoco reduced its emissions 3% below 1990 emissions in the first year of its program. For individuals, sixteen State Councils of Churches are participating in an Interfaith Global Climate Change Campaign. Under the campaign's auspices, individuals and families in Maine created carbon dioxide savings accounts equivalent to a 7% (the Kyoto target) reduction from 1990 personal and household emissions.

   In the spirit of our key findings, these bottom-up initiatives act both locally and globally. More important, they create the interest and commitment that presage needed political change.

REFERENCES

Appalachian State University Department of Geography and Planning. 1999. *North Carolina's Sensible Greenhouse Gas Reduction Strategies*. Boone, NC: Appalachian State University.

Harrington, J. Jr., D. Goodin, and K. Hilbert. 1997. Global Change in Local Places: Satellite Reflectance Data for Southwest Kansas. *Papers and Proceedings of Applied Geography Conferences*, **20**: 43–7.

Kasperson, J. X., and R. Kasperson. 2001. *International Workshop on Vulnerability and Global Environmental Change, 17–19 May 2001: A Workshop Summary*. Stockholm: Stockholm Environment Institute.

Kates, R. W., M. W. Mayfield, R. D. Torrie, and B. Witcher. 1998. Methods for Estimating Greenhouse Gas Emissions from Local Places. *Local Environment*, **3** (3): 279–97.

Lineback, N. G., T. Dellinger, L. F. Shienvold, B. Witcher, A. Reynolds, and L. E. Brown. 1999. Industrial Greenhouse Gas Emissions: Does CO_2 From Combusting of Biomass Residue Really Matter? *Climate Research*, **13**: 221–9.

Loretti, C. P., W. F. Wescott, and M. A. Eisenberg. 2000. *An Overview of Greenhouse Gas Emissions Inventory Issues*. Arlington, VA: Pew Center on Global Climate Change.

Margolick, M., and D. Russell. 2001. *Corporate Greenhouse Gas Reduction Targets*. Arlington, VA: Pew Center on Global Climate Change.

Soulé, P. T., and J. L. DeHart. 1998. Assessing IPAT Using Production- and Consumption-Based Measures of I. *Social Science Quarterly*, **79** (4): 54–765.

United States Global Change Research Program. 2000. *Climate Change Impacts on the United States: The Potential Consequences of Climate Variability and Change*. Cambridge: Cambridge University Press.

Wilbanks, T. J. 2002. Scaling Issues in Integrated Assessments of Climate Change. In J. Rotmans and M. van Asselt, eds. *Scaling Issues in Integrated Assessment*. Lisse: Swets and Zeitlinger.

Wilbanks, T. J., and R. W. Kates. 1999. Global Change in Local Places: How Scale Matters. *Climatic Change*, **43**: 601–28.

Index